Newton's Optical Writings

Masterworks of Discovery
Guided Studies of Great Texts in Science
Harvey M. Flaumenhaft, Series Editor

Newton's Optical Writings

A Guided Study

Dennis L. Sepper

Rutgers University Press

New Brunswick, New Jersey

Library of Congress Cataloging-in-Publication Data
Sepper, Dennis L.
    Newton's optical writings : a guided study / Dennis L. Sepper.
        p.   cm.
    Includes bibliographical references and index.
    ISBN 0-8135-2037-1 (cloth)   —ISBN 0-8135-2038-X (pbk.)
    1. Optics—Early works to 1800.   I. Title.
QC353.S46   1993
535—dc20   93-26703
                                                                                  CIP

British Cataloging-in Publication information available

To My Parents

Contents

Series Editor's Foreword

We often take for granted the terms, the premises, and the methods that prevail in our time and place. We take for granted, as the starting points for our own thinking, the outcomes of a process of thinking by our predecessors.

What happens is something like this: Questions are asked, and answers are given. These answers in turn provoke new questions, with their own answers. The new questions are built from the answers that were given to the old questions, but the old questions are now no longer asked. Foundations get covered over by what is built upon them.

Progress thus can lead to a kind of forgetfulness, making us less thoughtful in some ways than the people whom we go beyond. Hence, this series of guidebooks. The purpose of the series is to foster the reading of classic texts in science, including mathematics, so that readers will become more thoughtful by attending to the thinking that is out of sight but still at work in the achievements it has generated.

To be thoughtful human beings—to be thoughtful about what it is that makes us human—we need to read the record of the thinking that has shaped the world around us, and still shapes our minds as well. Scientific thinking is a fundamental part of this record—but a part that is read even less than the rest. It was not always so. Only recently has the prevalent division between the humanities and science come to be taken for granted. At one time, educated people read Euclid and Ptolemy along with Homer and Plato, whereas nowadays readers of Shakespeare and Rousseau rarely read Copernicus and Newton.

Often it is said that this is because books in science, unlike those in the humanities, simply become outdated: in science the past is held to be passé. But if science is essentially progressive, we can understand it only by seeing its progress *as* progress. This means that we ourselves must move through its progressive stages. We must think through the process of thought that has given us what we otherwise would thoughtlessly accept as given. By refusing to be the passive recipients of ready-made presuppositions and approaches, we can avoid becoming their prisoners. Only by actively taking part in discovery—only by engaging in its rediscovery ourselves—can we avoid both blind reaction against the scientific enterprise and blind submission to it.

When we combine the scientific quest for the roots of things with the humanistic endeavor to make the dead letter come alive in a thoughtful mind, then the past becomes a living source of wisdom that prepares us for the future—a more solid source of wisdom than vague attempts at being "interdisciplinary," which all too often merely provide an excuse for avoiding the study of scientific thought itself. The love of wisdom in its wholeness requires exploration of the sources of the things we take for granted—and this includes the thinking that has sorted out the various disciplines, making demarcations between fields as well as envisioning what is to be done within them.

Masterworks of Discovery has been developed to help nonspecialists gain access to formative writings in ancient and modern science. The volumes in this series are not books *about* thinkers and their thoughts. They are neither histories nor synopses that can take the place of the original works. The volumes are intended to provide guidance that will help nonspecialists to read for themselves the thinkers' own expressions of their thoughts. The volumes are products of a scholarship that is characterized by accessibility rather than originality, so that each guidebook can be read on its own without recourse to surveys of the history of science, or to accounts of the thinkers' lives and times, or to the latest scientific textbooks, or even to other volumes in the Masterworks of Discovery series

Although addressed to an audience that includes scientists as well as scholars in the humanities, the volumes in this series are meant to be readable by any intelligent person who has been exposed to the rudiments of high school science and mathematics. Individual guidebooks present carefully chosen selections of original scientific texts that are fundamental and thought provoking. Parts of books are presented when they will provide the most direct access to the heart of the matters that they treat. What is tacit in the texts is made explicit for readers who would otherwise be bewildered, or sometimes even be unaware that they should be bewildered. The guidebooks provide a generous supply of overviews, outlines, and diagrams. Besides explaining terminology that has fallen out of use or has changed its meaning, they also explain difficulties in the translation of certain terms and sentences. They alert readers to easily overlooked turning points in complicated arguments. They offer suggestions that help to show what is plausible in premises that may seem completely implausible at first glance. Important alternatives that are not considered in the text, but are not explicitly rejected either, are pointed out when this will help the reader think about what the text does explicitly consider. In order to provoke more thought about what are now the accepted teachings in science, the guidebooks bring forward questions about conclusions in the text that otherwise might merely be taken as confirmation of what is now prevailing doctrine.

Readers of these guidebooks will be unlikely to succumb to notions that reduce science to nothing more than an up-to-date body of concepts and facts and that reduce the humanities to frills left over in the world of learning after scientists have done the solid work. By their study of classic texts in science, readers of these guidebooks will be taking part in continuing education at the highest level. The education of a human being requires learning about the process by which the human race obtains its education—and there is no better way to do this than to read the writings of those master students who have been master teachers of the human race. These are the masterworks of discovery.

The writer most widely held to be the master of all the masters of modern discovery was Isaac Newton, and the most widely read of all of Newton's works was his treatment of light.

Light, that everyday feature of the natural world that illuminates all the things that we behold around us, had already in Greek antiquity become the subject of a mathematical science. When the study of Greek mathematical science was revived in early modern times, general progress was sought by a fundamental transformation in the study of nature: a universally mathematical physics was projected that would make masterworks of discovery the means of mastering nature. That project was put forward in the writings of the author who weaned Newton from his college-boy Aristotelianism. That author was Descartes, who reported that he had planned to go public initially with a book that presented his views on the whole wide world of nature in the form of a treatment of light. That book in which light provided the angle for a comprehensive view of the world, Descartes reported, he had been obliged to suppress in order to avoid a confrontation with the spiritual establishment, which to his surprise had intervened to block the progress of the Copernican transformation of astronomy.

Like the mathematical science of the movements of the lights that seem to circle us on high, the mathematical science of light was one of the first sciences to be transformed in modern times. But although both of those sciences had observational ingredients in their mathematical theorizing, the science of optics—unlike astronomy—involved experimental work as well. Both on its experimental side and theoretically, optics played a leading role in the development of early modern science, and it continued to do so later on, especially when light was reconceived as a wave in the nineteenth century and the quantum theory was developed in the twentieth.

The study of light was a very important part of the career of Isaac Newton. He invented a better telescope (of the reflecting type), and he developed techniques for the spectral analysis of light. But even more important than those more narrowly technical achievements was his achievement in transforming the whole science of optics. Although there may be disagreement about how accurate his results were, and about what interpretation should be given to those results, there cannot be any doubt that Newton made the science of light more rigorously experimental while at the same time demonstrating its mathematical principles. It was his great achievement in optics to reveal—by passing a narrow beam of white sunlight through a prism and getting a multicolored spectrum—that ordinary light is composed of many different kinds of light, each of them refrangible to a different degree and productive of a different color. But what may be most important about his optical work was its effect beyond the science of optics itself: Newton's optical work set new standards for experimental science generally, while suggesting new uses for light as a scientific instrument.

Newton in his optical writings arranged his experiments so artfully that the basic properties of light and the foundations of his theory of light seemed to demonstrate themselves. Newton's mode of presentation in optics became a model for eighteenth-century experimental investigators in many other fields. Many scientists since then have sought to follow Newton's lead in employing "crucial" experiments.

Newton's broader influence was not confined to method, however. Newton also led the way toward using the theoretical principles and experimental techniques of optics to probe the inner constitution of matter. He suggested that light might be made a means of uncovering nature's most deeply hidden mysteries, and thus that students of his optical writings might bring to light the fundamental laws that govern multifarious physical processes.

Despite that greatness of ambition and achievement, Newton's optical writings are remarkably accessible. The arguments they present do not require a very advanced knowledge of mathematics, nor do they depend on very complicated experiments. Even when Newton does perform a complicated experiment—using several apertures, prisms, and lenses—his readers usually can perform instead a worthwhile simplified version. Those readers of Newton's own words who also perform for themselves some of Newton's own experiments can get first-hand experience of the delights and the difficulties of doing experimental work in science. These optical writings thus make it possible for Newton's readers to become, in fact, co-workers in one of the greatest of the masterworks of discovery.

Nonetheless, the rich and colorful spectacle that Newton spreads before his readers can bedazzle and confuse a novice who approaches it without a guide. Beginners in the study of Newton's optical writings will therefore be happy to have the help of Professor Sepper. Professor Sepper himself came to these writings in the course of examining the great controversy in which Goethe criticized Newton's account of color. Goethe's critique raised probing questions about the relationship between experiment and experience, and about the place of mathematical rhetoric in the interpretation of nature. The book that Professor Sepper wrote some years ago on Goethe's critique of Newton was wonderfully illuminating for readers seeking a philosophical understanding of the controversy about Newton's optical work. Now, for readers seeking access to Newton's optical work itself, Professor Sepper provides this guidebook as part of the Masterworks of Discovery series.

Harvey Flaumenhaft, Series Editor
St. John's College in Annapolis—April 1993

Preface and Acknowledgments

This guided study of the optics of Isaac Newton was undertaken in the conviction that a classic in the history of science is rarely a dead letter for being old, that it can remain alive for later ages in the ways that scientists (and even ordinary people) think and act. This holds not just for "facts" and "laws" that may have been discovered but also for the whys, the hows, and the whithers: why the subject matter is of interest; how it is conceived, delimited, and approached; in what directions investigation leads. These things often tend to be overlooked, by experts because their interests are determined by specialized research at a scientific frontier, so that they have no reason to look back to historical classics, and by amateurs because they often lack the scientific knowledge necessary to recognize the virtues of a work.

The virtues of a great work, whether it is literary or scientific, theoretical or practical, are embedded in the work's artfulness and details. (That there are such things as great works, and that their greatness is sufficiently evident to those who have examined them carefully, are assumed for the purposes of these remarks.) Many people have wanted to write a great poem; a much smaller number have actually achieved a certain degree of excellence; but only a few have had the fortune to compose truly great poems. For the most part these have come about not in a flash of inspiration but rather line by line, even word by word, until the poem said what it was possible to say and reverberated with significance. Because of this such poems not only stand up to intensive scrutiny, they positively invite it and reveal remarkable density and texture to those who take them to heart. Great scientific works are much the same. They are works of scientific art that say what it was possible to say and that reverberate with meaning, most of all for those who are prepared to listen and are willing to make the effort of following along the paths they lead.

This guide is intended most of all for those who have become interested enough to want to read some of Newton's science but who have little or no scientific, philosophical, or historical preparation. From the outset such a reader must realize that the whole truth is not contained here. The reader will, I hope, learn a great deal about Newton's optics; about its scientific, philosophical, and cultural background; and about the nature of experimental and theoretical science and the reasons for doing it. But this is only a beginning, more an orientation to a text than an exhaustion of its meanings. This guide will therefore have succeeded best if, after finishing it, the reader decides that it would be worthwhile to have another go at Newton, armed with fuller and more thorough helps. The select bibliography at the end will give this reader some indications about where to turn next.

I hope that readers who already have advanced knowledge will nevertheless judge this guide to have certain merits. It is possible that, as someone once wrote of another work, it has something to dissatisfy everyone. I have tried not to be tendentious, and I have striven to avoid burdening the commentary with discussions of scholarly literature. I do wish, however, to acknowledge here my deep indebtedness to past and current Newton scholarship, which to the expert will be evident throughout. And I strongly urge general readers to look into the writings of the authors cited in the bibliography, especially Johannes Lohne, Alan Shapiro, Richard Westfall, Henry Guerlac, Thomas Kuhn, I. B. Cohen, B. J. Dobbs, Zev Bechler, and all the other scholars, both living and dead, who have made Newton studies and the history of science the vigorous enterprises they are today.

The selections from Newton's writings include nearly the entire letter to the Royal Society of 6 February 1672 and excerpts from all three books of the *Opticks*, including a generous selection from the Queries of Book III. Providing the letter virtually whole will give the reader a chance to follow a piece of scientific work from beginning to end. It also makes it possible to interweave, in a natural and relatively unobtrusive way, the historical context of Newton's optical investigations. In any case, the letter is an excellent, compact introduction to the project that culminated in the *Opticks*. The limits of space made it impossible to include more than a sampling of the *Opticks*, but I believe the selection is representative and gives an ample sense of how far beyond optics Newton's talents and interests extended.

The selected texts follow the originals almost slavishly, including spelling and punctuation, except for corrections of obvious or important errors, which I have indicated in the endnotes, and for the use of commas to group digits in very large numbers. I have also modernized the long *s*. The conventions of late-seventeenth- and early-eighteenth-century English style do not present major difficulties for the reader, and they not only give the writing an authentic flavor but also sometimes produce subtleties of emphasis that modernization would obscure.

I have not included any of the fourth edition's references to Newton's optical lectures. In general I have omitted Newton's own specific references to the figures, and the bracketed references to figures in the quotations from Newton are located more for convenience than for fidelity to the original. The positioning of headings with following number (e.g., "Book I," "Part IV," "Experiment 6," "Observation 15," "Query 31") generally imitates the original, except for "Axiom __," "Definition," "Proposition __, Theorem __," and "Proposition __, Problem __," which in the original were centered on lines of their own; in order to economize on space, I have put them on the same line as the beginning of the text they head.

My commentary on the letter and on Book I of the *Opticks* focuses chiefly on optics proper (especially differential refrangibility according to color), on Newton's disputes with his contemporaries, and on his conception of the proper method for understanding nature. I treat Books II and III not merely as optical science but also as an extension and transformation of optics into an instrument for investigating the deep structures of physical nature. The commentary on the Queries does little more than open up for the student a few leading questions, but I believe that I have provided at least a starting point for deeper study.

I have chosen to alternate text with commentary for a simple pedagogical reason. Good arguments are well-made arguments. Their excellence resides in details and in the artfulness of execution. But these are the hardest things to follow, and if you ask beginning readers to absorb excessively long passages without commentary (or even only with footnotes), they will typically lose track of what they have read. But I also understand the need to let readers hear Newton's voice and not just my own; so where the major difficulties of a passage have been cleared by what has come before, I allow longer breath to the original text.

The figures are intended to be more illustrative than physically exact; in general they are not drawn to scale. All figures that appeared in the original texts have been redrawn rather than photographically reproduced for the sake of both consistency and clarity, but in every case I have tried to preserve the character of the original, even where it is misleading or erroneous.

Those who have looked at my study of Johann Wolfgang von Goethe's polemics against Newton's theory of color will doubtless find here themes raised in that book, but this guided study has been undertaken from a different perspective, dictated by the need to make Newton understandable to a wide audience. In neither case has it been my intention to be the partisan of a narrow point of view but rather of a text and what it is about. In any case, the greater a work is, the more it challenges one to open one's mind and take the measure of the preconceptions of one's own and one's age.

Since the texts are arguments, a major part of my task is to clarify what is being argued and who the audience is. This is more than a question of giving historical background, for good scientific writings—good writings pure and simple, for that matter—are always contemporary. My hope is that this guided study will help readers think about each step as the argument progresses and encourage them to examine the evidence for it.

As with all scientific works, there are many unresolved problems, and even errors, in Newton's optical writings. At least a few of them are important for understanding Newton's theory and its reception, so I have included discussion of these in the main commentary. There are others, however—for example, concerning whether or not some of Newton's descriptions of experimental phenomena are really accurate—that, although important to anyone who tries to follow Newton's arguments and experiments carefully, did not have a major historical impact. These I raise in the endnotes; they are distinguished in the text from other references by an asterisk. By reflecting on such issues the reader may gain further insights into such questions as the relationship between natural perception and scientific theorizing, or whether the historical debates might have taken alternative paths.

The best way to see what science is about is to do it, or at least to try. I therefore cannot urge readers too strongly to get hold of some basic prisms and lenses and try to reproduce at least a few of the experiments described in the text. (A phone call to the department of physics of any college or university will be sufficient to get names and phone numbers for suppliers of inexpensive optical equipment.) It is in fact easy to perform many of them, at least in a simplified way. It would be a shame for nonprofessional readers of Newton's optical writings to miss an opportunity that few other scientific writings provide: not merely to follow the argument but to do what the original investigator did and see what that investigator saw. By following along with prisms and lenses readers will discover a much richer and more living science than if they only read the words.

Now for some thanks. Harvey Flaumenhaft inadvertently picked the worst possible day for trying to interest me in this project—he called at the very moment my printer was spewing out the final version of *Goethe contra Newton*. I was more than ready to move on to other things that I had long put aside to finish that book. I am glad he persisted, not least since I discovered there were still a few things to learn from Newton. I hope that this book measures up to the ambitions of the series he heads.

I must express my deep gratitude to the students of my Spring 1990 course in the philosophy of science for being very willing and very capable participants in the experiment of trying out the first draft of this guided study. I have benefited from their reactions and thoughtful comments on nearly every page. A special thanks to Cathy Wegner for helping make sure that it reached them in usable form.

Research and writing were supported by National Endowment for the Humanities grant RH-20915-89, under the Guided Studies initiative of the Humanities, Science and Technology category, part of Interpretive Research in the Division of Research Programs. My deep thanks to all the people at the Endowment who work so hard to put a small budget to the best use, and especially to Elizabeth Arndt and Daniel Jones of the HST program, who offered frequent and always prompt assistance throughout the term of the grant. Thanks also to the University of Dallas for permitting adjustments to my schedule that facilitated writing and permitted the timely use of the first draft of this guide in the classroom.

I am grateful to the librarians of the University of Dallas for their constant help and support: Susan Kendall, Dr. Zary Shafa, Claudette Hagel, Dr. Harry Butler, Lely White, Yolanda Garcia, Larry Jo Worley, and of course the director, Nettie Baker. A very special thank you to Alice Puro, who as always was unfailingly helpful in securing hard-to-get books and articles. Thanks to the librarians of the University of Florida, Gainesville, the University of Wisconsin, Madison, and the University of Texas, Austin (in particular the Harry Ransom Humanities Center), for their assistance during my visits to their campuses.

Many thanks to Karen Reeds of the Rutgers University Press for helping speed the process of turning the manuscript into a book, and my gratitude to everyone else at Rutgers Press and Princeton Editorial Associates, especially Peter Strupp, for their efforts.

I wish to thank the Warden and Fellows of New College, Oxford, for the right to reproduce Newton's autograph sketch of a version of the experimentum crucis, a sketch that was the basis for the frontispiece to the second French edition of the *Opticks*. I thank also the Department of Special Collections, Memorial Library, University of Wisconsin, Madison, for permission to photograph and reproduce the frontispiece portrait of Newton from a Latin edition of the *Opticks* published in Lausanne and Geneva in 1740. (It was first used in the third edition of the *Principia mathematica philosophiae naturalis* and is based on a 1725 portrait of Newton at age 82 by John Vanderbank.)

Thanks to the members of the Physics Department of the University of Dallas: Father Benedict Monostori, Richard Olenick, Sally Hicks, and Barney Ricca. Father Ben helped me find my way around the optics lab, and Richard and Sally set up experiments at a moment's notice when there was a question of fact to be resolved. To Barney I owe a special debt of gratitude for his patience in preparing the figures from my often inadequate sketches.

I thank my parents, to whom this book is dedicated, for the many sacrifices they made for the sake of my education and the love they showed in encouraging me to pursue it. I hope that at least a few of their expectations have been fulfilled.

I express not just thanks but also a husband's and parent's love to my family for their support: to my wife, Kathleen Wellman, particularly for arranging things whenever deadlines had to be met (I hope to return the favor as she progresses on her next book); and to my children, Elizabeth and Matthew—who have this curious notion that writing books and articles is their parents' leisure activity—for bringing a different and most welcomely human kind of craziness into the daily routine.

Conventions

For the text of the letter to the Royal Society of London of 6 February 1672,[1] I have followed the photographic reproduction of the *Philosophical Transactions,* no. 80 (19 February 1671/72), pp. 3075–3087, found in *Isaac Newton's Papers and Letters on Natural Philosophy,* second edition, pp. 47–59. For the *Opticks* I have used the fourth edition (London, 1730).[2] Any divergences from the originals are indicated in the endnotes.

In the letter there is some inconsistency of spelling, which I have not corrected. Some nouns are capitalized always, others occasionally, as a kind of emphasis. Commas are much more frequent than in modern usage, with somewhat different principles of placement; our "For if you consider that . . ." becomes "For, if you consider, that. . . ." The possessive often is not marked by an apostrophe, so "the sun's light" becomes "the Suns Light." In the *Opticks* spelling and punctuation are more regular and more similar to modern usage, but most nouns of importance are capitalized. To words that are likely to be confusing I have added a gloss in brackets at the first occurrence. Other problems related to spelling or punctuation can often be relieved simply by reading the passage aloud

Brackets within Newton's text around a word or phrase indicate a brief gloss or alternative to a difficult word or phrase immediately preceding, and around a sentence or clause indicate a brief explanation or summary. Brackets are also used to indicate figures. I have used [sic] infrequently, chiefly in cases where otherwise the reader might strongly suspect a printing error in this guided study.

Reference numbers accompanied by asterisks indicate that the note contains a more extensive discussion of problems with Newton's argument.

Wherever Newton's original text begins with a paragraph indention I have used an indention as well. Short omissions within a paragraph of the original are indicated by dots of ellipsis and no paragraph break. If dots of ellipsis at the end of a paragraph are immediately followed by a new indented paragraph with no line spacing, they mark the omission of just the latter part of the first paragraph. A line of dots of ellipsis between consecutive paragraphs of Newton indicates my omission of one or more full paragraphs of the original. When the dots are used at the very end of a quotation, they ordinarily signify the omission of just the remainder of the current paragraph, but occasionally they signify the omission of one or more additional paragraphs preceding the next quoted passage (this can be determined from the context). When commentary separates excerpts that are continuous in the original, there is no special mark; when the following excerpt is not continuous with the preceding one, and if there is not otherwise any sign of omission according to the preceding conventions, it begins (after any paragraph indention) with dots of ellipsis.

Notes

1. Gregorian calendar. The Julian calendar date was 16 February 1671 (sometimes given as 1671/72; it was 1671 in England, which observed New Year on 1 March).

2. The second (1717), third (1721), and fourth (1730) editions of the *Opticks* are basically the same. The second edition included all thirty-one Queries for the first time. The third was a new typesetting based on the second. The fourth edition closely followed the third (the pagination is almost identical), with corrections that, according to the preface, were Newton's. Yet the fourth was prepared at least in part by another editor or editors who included footnote references to the recently published optical lectures of Newton. A few of the fourth edition's "improvements" introduced errors, and some of the spelling and punctuation changes produced tiny inconsistencies over the course of the entire work. The 1952 Dover edition makes a small number of additional changes in spelling and punctuation to the version of the fourth edition. Although there are still good reasons for preferring the fourth, any of these editions is, with very few passages excepted, perfectly reliable.

Part 1

Preliminaries

Chapter 1

An Introduction to Isaac Newton

Nature, and Nature's Laws lay hid in Night.
God said, *let Newton be!* and All was *Light.*
Alexander Pope

Isaac Newton was born in Woolsthorpe, Lincolnshire, a small town in east central England, about one hundred miles south of London, on Christmas day 1642, two months after the death of his father. When his mother remarried in 1645 she left Isaac there in the care of her mother. As a child he attended the King's School in nearby Grantham. In 1656 he was entrusted with the task of operating the family farm in Woolsthorpe, to which his mother had returned after being widowed a second time. But his uncle, apparently recognizing Isaac's talents, urged him to return to his studies, and after another year of preparation he enrolled, in June 1661, at his uncle's alma mater, Trinity College, Cambridge University. He received the bachelor's degree in January 1665 and undertook further studies until an outbreak of the plague caused the University to shut down the following autumn. Newton retreated to his home in Lincolnshire.

Legend has it that the period of this retreat to Woolsthorpe was his *annus mirabilis,* his miraculous year, during which he made major breakthroughs in optics, mathematics, and the science of motion that were to serve as the foundation for his later achievements. More likely he was already engaged in this work when the plague struck Cambridge, and since the study of mathematics and the natural sciences were scarcely touched upon in the university curriculum of his day, it was probably no disadvantage to Newton to be in Woolsthorpe. When he returned to Cambridge he was made a Fellow of Trinity College, which provided a stipend and entitled him to a permanent place in the College, and the next year he was awarded the Master of Arts degree. In 1669 his mentor in the mathematical sciences, Isaac Barrow (1630–1677), retired from the Lucasian Chair of Mathematics and Astronomy, and Newton replaced him. In the next years he published important contributions to optics, based on lectures given at Cambridge, in a series of letters addressed to the premier scientific society of Europe, the Royal Society of London; and by the late 1670s he was working on problems of motion that led to the issuance in 1687 of the *Principia mathematica philosophiae naturalis, Mathematical Principles of Natural Philosophy*. It laid down in mathematical form the fundamental principles and laws governing the motion of bodies subject to forces (in a word, *dynamics*) and in particular gave an explanation of the dynamics of the solar system.

Although he had an intensely private, even secretive, personality, Newton also took on various public responsibilities and made the acquaintance of influential people. He engaged in university politics, and in the aftermath of the Glorious Revolution of 1688, which deposed James II and brought William and Mary to the throne of England, he was one of the emissaries sent to represent the University's interests to Parliament. In 1696 he resigned from the University to become Warden of the Mint in London, where he contributed to a reform of the currency and prosecuted (and even oversaw the execution of) counterfeiters; in 1699 he was made Master of the Mint. He became a member of Parliament and was knighted. In 1703 he was elected President of the Royal Society, a position he held till his death. In 1704 he published the *Opticks,* a work that can be said to mark the beginning of the modern science of light and color and that provided his contemporaries with a paradigm for how experimental science ought to be done. Although he published no single work in mathematics whose stature matches the *Opticks* or the *Principia,* he established himself, along with the German G. W. Leibniz (1646–1716), as the inventor of the calculus and discovered many other important mathematical theorems and techniques— not to mention that the *Principia* itself is one of the great historical examples of applying sophisticated mathematics to nature.

Newton also conducted extensive research into alchemical and chemical matters (the distinction between the two was still relative), he tried to confirm that the ancient philosophers and the Biblical prophets had understood the fundamental laws of the physical universe, and he published works that coordinated the chronology of events in the Bible with what was known from secular history. These were not simply hobbies: both the alchemical investigations and his religious and theological pursuits are relevant to understanding the man and what he hoped for from his science. Newton died in London at the age of eighty-three, on 20 March 1726.

Newton lived in an age when the full scope of the project of reconceiving the natural world was becoming clear. To mention only the most familiar names: in 1642, the year of Newton's birth, Copernicus's heliocentric theory of the solar system was a century old; Kepler's discovery that planetary orbits are elliptical, the tantalizing geometrical and proportional regularities he found in the arrangement and motions of the planets, and his reformulation and solution of leading optical problems were hardly more than a generation old; Galileo's refoundation of the study of motion had been made public just a decade before; and Descartes's proposals for a new method and a new conception of the universe as a mechanical system, alongside his major advances in optics and mathematics, were just five years previous. Informal associations of collectors of curiosities, scholars, and natural investigators were consolidating into the first scientific societies, and the advantages of considering the phenomena of nature as consequences of the operation of material mechanisms were beginning to appear. There was not yet a distinctive word to describe the activities of those who gathered experiences and speculated about the things of nature; the term *scientist* had not yet been coined (it was not used in English until about 1840), so instead they were given the traditional designation of *natural philosophers*. Nevertheless, these and others like them were giving birth to the early phase of the modern sciences of nature. These sciences were not yet so firmly established that they could not be decisively shaped by a single person. The work of Isaac Newton shaped not one but many sciences, and in his lifetime he was virtually canonized as the greatest investigator of nature ever. He quickly became a cultural icon, a living symbol of science, and his name and reputation lent a prestige to the investigation of nature that continues to this day.

Chapter 2

Optics in the Age of Newton

Optics. The word is very old. It was coined in ancient Greece, where the science thus named was flourishing over two thousand years ago. We understand it as the study of light and vision, including such topics as light's tendency to travel in straight lines (rectilinear propagation), the laws of reflection in straight and curved mirrors, the refraction of light in transparent substances, diffraction by finely etched surfaces, the focusing system of the eye and of cameras, and so on. The Greeks actually had distinct terms for these matters, each proper to a different class of phenomena: for example, reflection, the return or "rebounding" of light from smooth or mirrored surfaces, was the subject matter of *catoptrics,* and refraction, the sharp bending of light when it passes from one transparent medium into another (as with the bent appearance of a partially submerged oar), was studied by *dioptrics.* The word *optics* itself referred specifically to the study of vision; it is derived from the Greek word for eye, *ops.*

A brief history of the sciences of light and vision in antiquity would note the following: No later than around 300 B.C. the Greek "opticians" had developed a mathematical theory of how the eye perceives objects. By the middle of the first century A.D. the theory of reflection from flat and curved mirrors had achieved a rigorous mathematical form (although three hundred years earlier the great Archimedes [ca. 287–212 B.C.] had already attained a sophisticated understanding of reflection), and refraction was being investigated (the fundamental laws governing it were not discovered until much later, however). The ancient Greeks and Romans also noted the existence of other optical phenomena, such as the iridescence of fish scales and bird feathers, without making them the objects of thorough study or categorizing them in ways that would be recognizable today.

Quick summaries can introduce distortions in perspective, however, since they tend to assume that the real value of past investigations into nature depends on how close they came to achieving results that we consider correct today. For us, the central concern of optics is light and its behavior; vision is of only secondary importance, insofar as the eye can be conceived as a physical system of apertures and lenses that delimit and refract light (as is evident from the structure and subject matter of modern optics textbooks). On the one hand, then, there is a well-developed theory of light that (1) presents one or more models of the nature of light, for example that it consists of particles (call them photons) or of electromagnetic radiation (call them waves), and, largely on the basis of the strengths of such models, (2) explains the phenomena of the transmission of light through transparent media (including empty space) and its interaction with matter. On the other hand, there is a science, or rather many sciences, that take up the question of seeing. The explanation of vision begins with the eye but also quickly goes beyond it. From the camera-like system of the eyeball we would pass on through the electrochemistry of the optical nerve system to the brain, where complex "computational" processes involving relevant brain locations ultimately produce, or at least issue in, *seeing*. The behavior of light is taken as well understood, the biochemistry of the eye and nerves as somewhat less, but still well enough so that the focus of research can shift to the hierarchical and parallel systems that begin in the architecture of the optic nerves and the relevant areas of the brain.

In the earliest phases of the study of light and vision in Western civilization, however, the eye was conceived more as the endpoint rather than as the beginning of the process, and the chief phenomena of light were still in need of explanation and even of precise description and characterization. The idea of the rectilinear propagation of light seems quite commonplace, but it took a stroke of genius for someone to determine that light travels in straight lines (the chief evidence of which is the sharpness of shadows cast by sunlight). But what does it mean to say light *travels?* Here one needs to be careful, since the answer depends on what light is thought to be. For example, the Greek philosopher Aristotle (384–323 B.C.) thought of light as the activation of the transparency of a dark medium (like water or air) that in turn allowed the visible properties of things, in particular color, to be conveyed to the eye; thus for him it would be more proper to say that it is the color properties of things that are conveyed along straight lines, not light.

Even those who thought that light itself was what traveled between visible object and eye did not all agree on the source of the light and its direction of travel. For modern theories and many ancient ones, physical light comes from the "external" world and enters the eye; it is *intro-mitted*. For example, the atomists thought that objects emitted particles that bore the traces of the object's appearance. This combined into a single cause the reasons for both linear propagation and the formation of images. But other theorists of Greek and Roman antiquity and the Islamic and Christian middle ages thought that the eye emitted (or *extramitted*) a ray of its own, the *visual ray,* and that this visual ray came into direct contact either with the object or (in some versions) with an external light coming from the object. In this scheme seeing was conceived as taking place "out there" in the visible world, into which the visual ray reached out from the viewer's eye. Supporters of the theory could point to various kinds of evidence, for example the glowing of animals' eyes in the dark and the fact that we see things as being in the world rather than as in our eyes or heads.

Today we understand that the eyes of a cat glow only by virtue of reflected light, so that the notion of a light or "fire" sent out by its eyes seems quite wrong and even quite amusing. But we should not be smug in judging previous ages. After all, many of the things the average person claims to know scientifically he or she really knows by hearsay (including having heard it in a science class). If we can claim to understand the real reason that cats' eyes glow in the dark, we can hardly do the same for the other item of evidence mentioned above, that things appear to be in the world rather than in our heads. At best we might say that our minds or brains "project" an external world constructed out of the impulses gathered by our sense organs and processed by higher neurological centers, but there is more than a little speculation involved in such an assertion. Explanations like this suggest that we think the right kind of answer to the question will come from some science other than optics. This helps show that the notion of what subject matter belongs to a particular science and what belongs elsewhere can change over time (or even that a science may change because what was previously included in it comes to be assigned elsewhere). In addition, we should remember that modern science can be viewed as largely hypothetical (hypothetico-deductive), so that there is the expectation that most of what we think scientifically today will eventually be replaced by better theories, and some of it will be abandoned as wrong.

Investigations of light and vision in antiquity embraced traditions that did not so much exclude one another as emphasize different aspects of the subject. Euclid (flourished ca. 300 B.C.) advanced a chiefly mathematical perspective based on the straight-line propagation of light. Among the ancient atomists (e.g., Democritus [ca. 460–371 B.C.] and Lucretius [ca. 96–55 B.C.]) there was an emphasis on light as a physical entity. Ptolemy (ca. A.D. 100–170) combined the mathematical approach with perceptual, psychological, and philosophical matters. For the ancient physicians (e.g., Galen [ca. 130–200]) the emphasis was understandably on the nature of the eye and its operations. In addition, philosophical and psychological approaches, especially Aristotle's, continued to be cultivated and had a powerful influence on mathematical, physical, and medical optics. Although with the collapse of the Western Roman Empire in the fifth century A.D. this scientific and philosophical knowledge was rapidly lost to Europe, the different approaches spawned further interactions and developments over the centuries in the territories under Islamic rule in the Arabic and Persian middle ages;[1] but no single theory or combination of theories came to dominate the understanding of light and vision.

In the Latin High Middle Ages (beginning in the twelfth century) and continuing into the early Renaissance the achievements of the past were slowly reappropriated in the West until, in the late sixteenth and early seventeenth centuries, the sciences of light and vision became recognizably modern. A turning point occurred in the work of Johannes Kepler (1571–1630). Kepler brought about a synthesis between the mathematical and the eye-centered approaches by providing an explanation of the formation of images on the retina, an explanation that to us is recognizably "correct." Its fundamental principle is this: light rays are propagated from every point of an illuminated object in every available direction. When an eye is placed in the path of such rays, the pupil admits a circular cross section of the light into its interior. This light is refracted by the lens of the eye so that all the rays coming from any given point on the object converge on a single point of the retina. The result: on the retina there is reconstituted, point by point, an accurate though inverted image of the object that originally reflected (or emitted) rays in all directions.

In the generation after Kepler both Willebrord Snel (1580–1626) and René Descartes (1596–1650) discovered a geometrical or trigonometric regularity in the refraction of light, a regularity that is now usually called

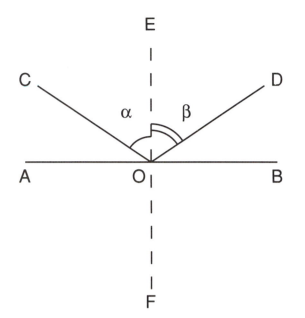

Fig. 2. AOB is a reflecting surface; EOF is an imaginary perpendicular intersecting it at O. Ray CO is reflected into path OD, so that the angle of incidence (COE or α) is equal to the angle of reflection (EOD or β).

the sine law of refraction.[2] To understand this principle, it will help to think first about reflection. Reflection (the simplest case is a flat mirror) is easily explained, at least in principle. At the point where the ray encounters the reflecting surface it is turned back in such a way that the angle of encounter (the *angle of incidence*) is equal to the angle of departure (the *angle of reflection*). In order to standardize terminology and descriptions, we will henceforth assume that the angles are measured from an imaginary line that is drawn perpendicular to the reflecting surface at the point of incidence (see Fig. 2). Imagining such a perpendicular in place, we can specify one other fact about reflection: it occurs in the geometrical plane determined by this perpendicular and the incident ray. If you think of the ray of light as a ball bouncing off a smooth floor, this means that from the perspective of the thrower the ball does not take a crazy bounce to one side or the other but remains in the plane perpendicular to the floor.

Refraction, the passage of a ray from one kind of transparent medium (like air) into another (like glass), is more complicated because the relationship between the angle of incidence on the refracting surface and

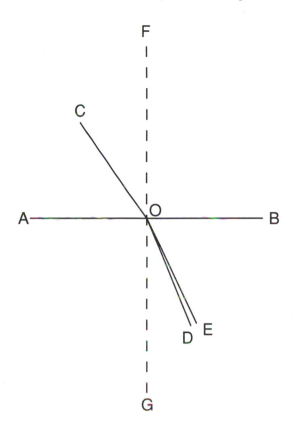

Fig. 3. If AOB is the interface between air (above) and water (below), ray CO will be refracted into OE; but if the lower material is glass, the refraction of the same ray CO will produce a path (OD) closer to the perpendicular FG.

the angle of departure or refraction within the second medium is not at all obvious, and since we have no experience of balls being refracted through floors the model we have just used to picture reflection doesn't get us anywhere. The ray does indeed remain in the same plane (determined by the perpendicular to the surface at the point of incidence and the line traveled by the incident ray), but now it enters the second medium instead of being driven back, and it changes its direction to a degree that changes with the material of the second medium.

For example, light that goes from air into water is turned out of its original direction less than light going at the same angle from air into glass (see Fig. 3). Although it seems in general that the degree of refraction increases when the second material is denser (so that diamond refracts to a greater degree than glass, and glass more than plastic), there is no strict proportion that holds between density and refraction.[3] Moreover, as the angle of incidence increases (that is, the farther the incident ray is from being perpendicular to the surface) so does the degree to which the ray is refracted, but finding a formula to describe this is not easy (unless you already know it!). Before Snel and Descartes other investigators had suggested various geometrical and trigonometric relationships, but these two were the first to give the exact relationship and to provide a rationale for why it holds. Their answer was mathematically equivalent to the law of sines (see Fig. 4): given a surface at which refraction takes place (that is, an *interface* where one transparent medium adjoins another), if AO and CO are any two rays incident on the interface at different angles, then the sine of the angle of incidence of the first (abbreviated "sin i_1") divided by the sine of its corresponding angle of refraction (sin r_1) will equal the sine of the angle of incidence of the other (sin i_2) divided by the sine of its angle of refraction (sin r_2). That is,

$$\sin i_1/\sin r_1 = \sin i_2/\sin r_2$$

The law of sines has a very handy consequence: if you measure the angles of incidence and refraction for a single ray going from material 1 into material 2, you are in effect determining the relationship of the angles for every other conceivable ray refracted from 1 into 2 at the same interface, because the sine of the incidence angle divided by the sine of the refraction angle gives a number that is the same for all rays. From that number (call it n) and any arbitrarily chosen angle of incidence, one can calculate the expected angle of refraction. The number n is called the *relative index of refraction*—relative, that is, to the interface of the two transparent materials being used (one of which is usually the air). By extension, one can define an *absolute index of refraction* for a material as the value of n for rays entering the material from the vacuum of empty space. For practical purposes, this absolute index differs very little from the relative index for rays entering a material from the earth's atmosphere.

There are two other important things about refraction to point out here, the *reversibility of ray paths* and the phenomenon of *total internal reflection* (and the related notion of the *critical angle* of incidence). First, for both reflection and refraction the paths traced out by rays are revers-

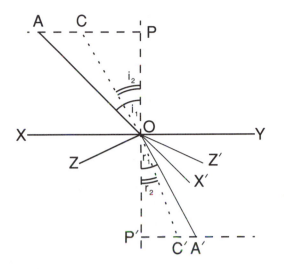

Fig. 4. XOY is the interface between two transparent media, with the lower one (say glass) being more optically dense than the upper (say air), POP' is a perpendicular to it, and AO and CO are two rays incident at different angles (i_1 = AOP and i_2 = COP, respectively). They are refracted into rays OA' (at angle r_1 = A'OP') and OC' (at angle r_2 = C'OP'), respectively. The law of sines says that for any two such rays,

$$\sin i_1/\sin i_2 = \sin r_1/\sin r_2$$

OX', OZ', and OZ illustrate the critical angle and the phenomenon of total internal reflection. OX' is the path that a ray very nearly parallel to XY would take if it entered the lower medium at O; for refraction from the upper into the lower medium no ray will have a larger angle of refraction (like OZ'). If Z'O is the path of a ray coming from below that strikes the interface at O, it will not be refracted into the upper, less optically dense medium but instead be totally reflected along OZ.

ible. This means that if at some point in the ray's path you placed a mirror to reverse the direction 180° (thus the ray would head back along the line it had just followed), it would trace out exactly the path it had taken from the beginning of the diagram, thereby reversing all the refractions and reflections that had taken place. In Figure 4, for example, the ray that travels from A to O and then is refracted toward A' would, if it started out from A' toward O, be refracted toward A. In the first case the incident angle is AOP and the refracted angle A'OP', whereas in the second case it is the other way around. If *n* indicates the relative index of refraction in the first case, $1/n$ is the relative index in the second. This makes clear that the relative index of refraction between two materials depends on the order in which they are taken; in fact, the inverse relationship between the two indexes is simply a mathematical way of expressing the reversibility of refraction.

Second, to examine total internal reflection and the critical angle, consider, again in Fig. 4, a ray going from air (above) into glass (below) at point O of the interface. If the ray is perpendicular (like PO), it will suffer no deviation from its path and so go straight on to P′. If the angle of incidence is increased, that is, if the ray is made less perpendicular, such as CO and AO, the angle of refraction will increase. As the ray entering at O becomes nearly parallel to the interface—that is, when it "grazes" the interface along XO—the angle of refraction will reach its maximum (X′OP′). So no ray entering the glass at O will ever take a path like OZ′, which is more inclined to the perpendicular than OX′.

Recalling that refraction is reversible, however, we recognize that a ray going from Z′ to O cannot be refracted into the air—otherwise reversibility would ensure that OZ′ was a possible path for some ray entering from above, contrary to what we already know. So what happens to a ray traveling in the glass from Z′ to O? It is reflected, not refracted, just as though the interface XOY had become a perfect mirror! The light ray therefore stays in the lower medium, along OZ, and is said to undergo *total internal reflection*. The angle X′OP′, the greatest angle of refraction for any ray entering from the less dense medium, is called the *critical angle,* beyond which there are no admissible paths for a refracted ray.

You can easily experience this mirror-like effect in a right-angled prism with two 45° angles. It happens that all kinds of glass have critical angles less than 45°. A ray entering perpendicularly through one of the two short sides will be incident on the interface opposite at exactly 45°, more than the critical angle, and therefore will be reflected so as to exit perpendicularly through the other short side. The critical angle and total internal reflection have also been experienced by most swimmers. If while submerged you look up, you see the sky occupying a relatively small disk-like area above your head; and if there are fish swimming nearby you can often see their images reflected at the surface of the water.

With the work of Kepler, Snel, and Descartes, the sciences of light and vision reached a new level of perfection. The question of where in the eye the image was formed had received a definitive answer—it was neither the cornea nor the lens, as had been anciently thought, but the back of the eye, the retina. The question of vision was therefore either answered (if vision amounted to the formation of an image in the eye) or reduced to the question of what happens once the inverted retinal image is formed. The image itself resulted solely from what happened to rays of light entering at the pupil. The behavior of light in both reflection and refraction could now be described in exact mathematics—two types of phenomena that, at least at first glance, seem to cover exhaustively the behavior of light in the physical world.[4] Moreover, through mathematical representation the behavior of beams of light could be analyzed into independent smaller parts (represented as mathematical rays). Although this was not the first time in history that light was conceived as separable into components, it was now clear that the exact determination of images required accounting for the behavior of all the individual rays and taking the sum.[5] This in turn suggested raising the question of what these rays were and, by corollary, whether there were infinitely many of them, as would be the case if light were infinitely divisible, or only finitely many, if there were in fact some smallest part of light.

Seventeenth-century developments thus brought about a division of scientific labor: what more (if anything) was needed to explain vision could be left to those who know what happens between the retina and the brain (presumably physicians), whereas other investigators, who eventually would come to be called physicists, were responsible for examining the behavior of light in the world outside the living body. The dividing point between the two realms was the retinal image.

Henceforth the term *optics* named not the study of vision but a science that was both physical, in the modern sense, and mathematical, as it had been since Greek antiquity, and that had as its object physical light understood not as something emitted by the eye but as produced by luminous objects in the physical world. Kepler, Snel, and Descartes had enriched the ancient mathematical approach with powerful techniques for analyzing light into individual rays or paths governed by determinate laws of reflection and refraction. Alongside the mathematical treatment describing ray paths, there was also a developing interest in the physical things that traveled those paths. This physical side of optics received an impetus especially from people like (on the one hand) Descartes, who interpreted light as a special impulse or pressure exerted in very fine, invisible matter, and (on the other hand) the French writer who popularized atomism, Pierre Gassendi (1592–1655), and, at one point in his career, the English philosopher Thomas Hobbes (1588–1679), who advanced a corpuscular optics, that is, a theory based on light conceived as consisting of very many tiny bodies flying in straight lines at incredible speed. These schools of thought began a debate that would continue for centuries: was light a pressure, wave, or impulse phenomenon, or was it a multitude of small bodies traveling at extremely high velocities?

Newton was quite aware of the new traditions in optics, as can be seen from a notebook he began while pursuing his bachelor's degree at Cambridge University, probably in 1664.[6] It shows that he was actively reading the chief representatives of the new mechanical and atomistic philosophies, including Descartes, Hobbes, Gassendi, Walter Charleton (1620–1707; he was an English follower of Gassendi), and Robert Boyle (1627–1691). It shows as well that he was constantly questioning, criticizing, and evaluating what he found in them, and that he was intent on working out the consequences of theories and devising experiments to test them. Other notes written a little later, probably in 1665–1666, show that Newton had progressed rapidly in his knowledge of optics. They describe prism experiments recognizably like those of his mature optical writings, covering reflection, refraction, and even the phenomena of colors produced by thin films (as in soap bubbles or dirty oil floating on water) that he attributed to a property he called *inflection*. They also contain fundamental principles of his theory of the differential refraction of rays according to color.

Most of these experiments, theories, and questions Newton gathered under the thematic heading "Of Colors," and they include phenomena and speculations regarding the eye and its functioning in the perception of images and colors. One thing they show is that even at this early stage of his work Newton rejected the idea that light could be a wave, pulse, or pressure phenomenon and that he conceived of rays as being little spheres or *globuli*. They also show that Newton's "optical speculations began in the physiological context of color perception, and he moved on to consider the phenomena of colors as seen through the refracted light of a prism."[7] Descartes had also been intrigued by the physiological and psychological questions as well as the physical ones some forty to fifty years earlier, and there is a strong case to be made that it was precisely this influence that shaped Newton's earliest optical research; but it is important to recall that these questions were also central in the tradition that preceded Descartes.

Within a few years of these experiments Newton became the Lucasian Professor of Mathematics of Trinity College. In 1670–1672 Newton used the position to present, as had Barrow, lectures in optics,[8] with the difference that Newton discoursed not just about geometric optics, which treats light in terms of mathematical lines and rays, but also about physical optics and color. The lectures doubtless exceeded the understanding of his students, who would have known little more than a bit of arithmetic and geometry, and probably none of whom would have ever encountered a mathematized science of nature before. The lectures would probably have been without effect had Newton's researches not led him to construct a new type of reflecting telescope.[9] Apparently at the urging of others, Newton submitted the telescope to a recently founded society dedicated to the accumulation of knowledge about nature with the promise that he would shortly provide a theoretical justification for its design. The justification was contained in a letter to the Royal Society of London dated 6 February 1672, a letter that, in summarizing the results of his lectures, set the unknown Fellow of Trinity College on a path that would lead to controversy and, ultimately, world fame.

Part 2

The Early Optics

Chapter 3

Newton's First Publication, the 1672 Letter

In January 1672 Newton wrote to Henry Oldenbourg (1615–1677), secretary of the Royal Society of London, to inform him that he would soon receive a letter describing "the oddest if not the most considerable detection which hath been made in the operations of nature,"[1] one that had led Newton to the invention of the reflecting telescope that was already in the possession of the Society. That letter, dated 6 February 1672, is one of the masterpieces of scientific literature, and in an important sense it is the first great specimen of what has since become the standard genre of scientific writing, the journal article. At least on the surface, it seems hardly to need any commentary even after more than 300 years, so clear and to the point is its prose. The tone is set already in the opening paragraph:

To perform my late promise to you, I shall without further ceremony acquaint you, that in the beginning of the Year 1666 (at which time I applyed my self to the grinding of Optick glasses of other figures than *Spherical*,) I procured me a Triangular glass-Prisme, to try therewith the celebrated *Phaenomena* of *Colours*. And in order thereto having darkened my chamber, and made a small hole in my window-shuts [chutters], to let in a convenient quantity of the Suns light, I placed my Prisme at his entrance, that it might be thereby refracted to the opposite wall. It was at first a very pleasing divertisement [diversion], to view the vivid and intense colours produced thereby; but after a while applying my self to consider them more circumspectly, I became surprised to see them in an *oblong* form; which, according to the received laws of Refraction, I expected should have been *circular*.

They were terminated at the sides with streight [straight] lines, but at the ends, the decay of light was so gradual, that it was difficult to determine justly, what was their figure; yet they seemed *semicircular* [see Fig. 5].

Fig. 5

Newton is describing here the paradigm case of producing colors by the refraction of sunlight entering through a small aperture into an otherwise darkened room or chamber (*camera obscura* in Latin, from which we get "camera" for the photographic instrument). In practice it is not even necessary to completely darken the room to get a good spectrum of colors; all that is needed is to make sure it is less bright inside than outside. For example, stand by a window facing the sun and hold a prism in the light: with the refracting angle pointing downward the light will be refracted upward, and high on the wall or on the ceiling you will see the classic spectrum. (One thing you will probably notice is that Newton's description of the shape of the spectrum is a simplification.)[2*] If you rotate the prism back and forth around its long axis (Fig. 6), you can make the spectrum change position. With the refracting angle downward, when the angle of incidence on the first face is equal to the angle of refraction on the second, the spectrum will be as low on the wall as possible. Whether you rotate the prism one way or the other, the spectrum will tend to rise. For the moment, let us call this the preferred orientation of the prism.

Newton was entertained and charmed by the colors he saw, just as we are today. But why did he go to the trouble of procuring the glass prism

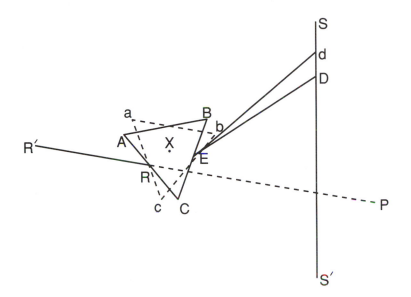

Fig. 6. ABC is one end of the prism (a cross section), and X is the imaginary central axis running the length of the glass. By rotating the prism around X the orientation can be changed, as from ABC to abc. This rotation changes the position of the spectrum on the screen SS'. Let R'R be an incident monochromatic (homogeneal) ray for which the index of refraction is 1.5 (from air to glass) that is refracted to position D for orientation ABC and to d for orientation abc. If ABC is the position of symmetry (so that the angle of incidence at face AC is equal to the angle of refraction at face CB), at that position ray R'R is minimally deviated from its original path; any other position, such as abc, will refract the light to a higher point on the screen.

in the first place? He says he had been trying to grind lenses of nonspherical shape; presumably he hoped to learn something about these lenses by studying prisms.

First, what is the significance of nonspherical lenses? Earlier in the seventeenth century it had been shown (by René Descartes) that lenses whose curved surfaces are segments of spheres[3] cannot produce perfectly focused images because they do not bring to a single point all the rays coming from a given point of the object being imaged; that is, the spherical shape produces image distortion. It had further been shown that certain other shapes, in particular elliptical and hyperbolic curves, could in theory be used to overcome this defect and produce a perfectly

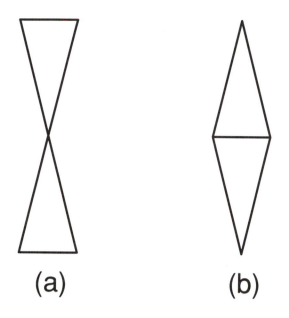

(a)   (b)

Fig. 7. Approximating lenses with prisms. (a) Two prisms vertex to vertex resemble a double-concave (biconcave) lens. (b) Two prisms base to base resemble a double-convex (biconvex) lens.

focused image. The problem was in manufacturing such a lens, a process that requires special machinery and techniques. (Ordinary lens-grinding techniques produce spherical lenses.) It was widely believed that a significant improvement in telescopes and microscopes had to await the invention of reliable production techniques for such nonspherical lenses.

But if lens curvature is the issue, why bother with prisms, which have triangular cross sections and flat faces? Because a prism can be conceived, at least schematically, as an element or part of a lens; the lens can be conceived as the result of fitting together prisms. Two prisms put vertex to vertex resemble a primitive concave lens, whereas set base to base they resemble a convex lens (see Figs. 7a and 7b). By imagining a lens divided into many sections as in Fig. 8, you can see how it might be approximated using prisms with triangular and trapezoidal cross sections.

The upshot is that studying what a prism does to light is like studying what a small part of a lens does to light. It allows you to isolate the behavior of a part of the light that goes into the formation of a larger image without the complication and overlap produced by the simul-

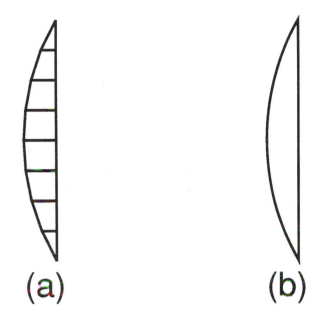

Fig. 8. Multiple prisms of triangular and trapezoidal cross section (a) can be used to approximate the curvature of a lens (b).

taneous refraction of all the other parts of the light at different angles because of the curvature of the lens. Using a prism therefore allows for both theoretical and practical simplification in studying the effects of a lens.[4*]

The last preliminary is "the received laws of refraction," that is, the sine law of refraction, which led Newton to expect a circular rather than an elongated image. He does not explain how he arrived at the conviction that the spectrum should be circular. In fact, if a unique value for the sine law governed the refraction, for any randomly chosen angle of incidence on the first face of the prism the image would *not* be perfectly circular but would be somewhat elongated instead (though not the five times that Newton describes). Today the calculation of the result for any randomly taken angle of refraction is simple, and one could easily write a computer program to display the results almost instantly, but lacking such powerful instruments the calculation or geometric construction can be laborious. Fortunately one does not need to make calculations for the preferred orientation of the prism, as Newton showed in his optical lectures. At the preferred orientation (technically known as the *position of minimum deviation*), any rays that are parallel to one another before refraction will be parallel after as well, and any rays incident at slightly different angles will have their orientation changed only slightly. Since even the extreme parts of the narrow beam of sunlight admitted into the darkened room are inclined to one another no more than about half a degree, the circular shape of the image will be maintained at the preferred position of minimum deviation. This is a relatively simple consequence of the geometry of the prism.

The position of minimum deviation turns out to be a position of symmetry: the angle of incidence on the first face, where the light enters the prism, is equal to the angle of refraction at the second face, where the light leaves it (see position ABC in Fig. 6). If one rotated the prism so that the beam entered at angles further and further from the angle of minimum deviation, the image would suffer an increasing elongation.[5*]

Unfortunately Newton did not explain these things to his intended audience, the members of the Royal Society of London, relatively few of whom knew much more about the laws and mathematics of optics than had his students at Cambridge. In the initial reaction to the letter there was some misunderstanding of this and other "elementary" principles of optics; if Newton genuinely expected everyone to grasp these things, he sadly overestimated their abilities. Moreover, if at the time of the prism experiment that Newton is describing he had already derived such mathematical results, this means that he was not just playing with the prism (as the letter suggests) but already in possession of sophisticated knowledge about the phenomena it produces. Perhaps, then, the letter is not so much the literal account of events as a dramatic reconstruction of them.[6]

Still, there are ways of looking at the situation that would make it plausible that things happened more or less as narrated. First, Newton assisted Isaac Barrow at some stage in the preparation of the latter's geometrical optics and so probably knew a good deal of technical theory by 1666. In that case the prism experiments would have been intended to let Newton see in fact what he knew in theory; Newton simply erred in assuming that others knew as much basic theory as he did. Second, the laws of refraction as they were presented in the seventeenth century did not systematically discriminate between the behavior of the beam as a whole and the components of that beam. Even without engaging in complex calculations or geometrical proofs, one could think of the whole beam as behaving more or less like a single mathematical ray. This way of looking at refraction is reinforced by the expectation that, if all the parts of the beam are basically traveling parallel to one another before refraction (as is the case with rays coming from the sun, ninety-three million miles away), they should continue to be basically parallel afterward. Thus a circular beam of sunlight before refraction should still be more or less circular afterward (as long as the screen or wall on which the image is cast is perpendicular to the beam's path; by holding the screen at an extremely oblique angle to the beam, one will of course elongate the image). If the elongation of the image is sufficiently great in the situation Newton has set up (a narrow beam of sunlight coming into a dark room), there will be need for an explanation. Newton continues his narrative precisely by puzzling over this odd phenomenon and its possible causes.

Comparing the length of this coloured Spectrum[7] with its breadth, I found it about five times greater; a disproportion so extravagant, that it excited me to a more than ordinary curiosity of examining, from whence it might proceed. I could scarce think, that the various *Thickness* of the glass, or the termination with shadow or darkness, could have any Influence on light to produce such an effect; yet I thought it not amiss, first to examine those circumstances, and so tryed, what would happen by transmitting light through parts of the glass of divers thicknesses, or through holes in the window of divers bignesses, or by setting the Prisme without so, that the light might pass through it, and be refracted before it was torminated by the hole: But I found none of those circumstances material. The fashion of the colours was in all these cases the same.

Later Newton will resort to exact measurements; for now he is assessing things in a rough and ready way, and he is looking for a possible explanation not in the precision of measurement and theoretical calculation but by entertaining some hypotheses about what is behind the extraordinary elongation. The hypotheses he considers are not pulled out of his fingertips. He had undoubtedly encountered some of them in his reading about the phenomena of light and color. For example, Marko Antonije Dominis (1560–1626) had suggested that this spectrum came about because in a triangular glass prism one side of the beam had to traverse a slightly longer path in the glass than the side opposite (and thus was more obscured by the slight opacity of the otherwise transparent material of the prism). A whole series of natural philosophers had argued that the colors of refraction are produced by some disturbance or weakening of the light, beginning at the edges of the beam. At least a few

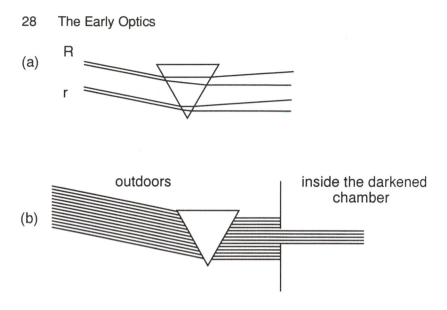

Fig. 9. (a) Newton's first variation: passing rays through different thicknesses of the prism, using rays R and r. (b) Newton's third variation: placing the prism outside the windowshut.

interpreted this as evidence that colors were a mixture of light and darkness (in antiquity Aristotle had argued that colors are produced by the interaction of light and darkness).

To test these hypotheses Newton used a very simple technique: introduce variations of the experiment so that if the hypotheses are true the outcome (the appearance of the spectrum) should be changed. So he refracts the light both near the prism's refracting angle, where the light is in the glass for only a short space, and then near the base, where there is a longer path through the glass (Fig. 9a). The spectrum remains unchanged, so the hypothesis concerning the length of the path in the prism contributes nothing toward an explanation. He tries larger and smaller holes, without (he says) any important new results. Then, instead of refracting the light after it enters the room through the hole in the shutter, he (or an assistant) takes the prism outside, so that the light is refracted before it passes through the hole (Fig. 9b). Again he finds no material change in the outcome, so this hypothesis is also rejected.[8*]

Then I suspected, whether by any *unevenness* in the glass, or other contingent irregularity, these colours might be thus dilated. And to try this, I took another Prisme like the former, and so placed it, that the light, passing through them both,

Fig. 10. The first two-prism experiment. Identical prisms, the second rotated 180°, cancel one another's effects, so that ray AB ends up on the parallel path CD.

might be refracted contrary ways, and so by the latter returned into that course, from which the former had diverted it [Fig. 10]. For, by this means I thought, the *regular* effects of the first Prisme would be destroyed by the second Prisme, but the *irregular* ones more augmented, by the multiplicity of refractions. The event was, that the light, which by the first Prisme was diffused into an *oblong* form, was by the second reduced into an *orbicular* [circular] one with as much regularity, as when it did not at all pass through them. So that, what ever was the cause of that length, 'twas not any contingent irregularity.

Newton is entertaining another hypothesis: perhaps the colors and the elongation are produced by irregularities in the glass that scatter the light.[9] In contrast to modern prisms, the best ones that Newton would have been able to obtain (or make himself) would have been relatively small, imperfectly transparent (with bubbles, streaks, and other imperfections, including traces of color), and at least a bit irregular in shape (in particular, the faces would have rarely been perfectly flat planes). Thus irregularities were common, so if they turned out to be the cause of colors and elongation then multiple refractions should make the phenomenon even more conspicuous. In addition, if irregularities cause refraction, then the process must be more or less random; it should be impossible for an additional refraction to undo the randomness or irregularity produced by a preceding refraction. Two very similar prisms, the second inverted with respect to the first, produce four refractions (since the light passes through two interfaces with each prism); the result Newton obtains is a colorless, circular image, as though there had been no refraction at all.[10*] Therefore, irregularities cannot be responsible for the elongation of the image.

Newton has now tested three hypotheses about refraction, but none of them explains the elongation. Frustrated in his search for a quick answer from existing hypotheses, Newton turns to a more precise examination

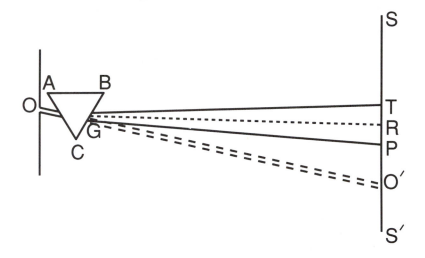

Fig. 11. Newton's measurements. Aperture O is ¼ inch in diameter; the distance from O (or prism ABC) to the wall SS′ is about 22 feet; the spectrum PT is 13¼ inches long (and 2⅝ inches wide); the angle C of the prism is 63° 12′; the angle between the path that the rays would have followed had there been no refraction (along OO′) and the path of the middle ray of the refracted light (the dashed line going to R) is 44° 56′; and the initial angle of incidence (at face AC) and the final angle of refraction (at BC) are both 54° 4′.

of the spectrum and the experiment that produces it, in particular the precise angles at which the rays are incident.

I then proceeded to examin more critically, what might be effected by the difference of the incidence of Rays coming from divers parts of the Sun; and to that end, measured the several lines and angles, belonging to the Image [Fig. 11]. Its distance from the hole or Prisme was 22 foot; its utmost length 13 1/4 inches; its breadth 2 5/8; the diameter of the hole 1/4 of an inch; the angle, with the Rays, tending towards the middle of the image, made with those lines, in which they would have proceeded without refraction, was 44 deg. 56′.[11] And the vertical Angle of the Prisme, 63 deg. 12′. Also the Refractions on both sides the Prisme, that is, of the Incident, and Emergent Rays, were as near, as I could make them, equal, and consequently about 54 deg. 4′. And the Rays fell perpendicularly upon the wall. Now subducting [subtracting] the diameter of the hole from the length and breadth of the Image, there remains 13 Inches the length, and 2 3/8 the breadth, comprehended by those Rays, which passed through the center of the said hole, and consequently the angle of the hole, which that breadth subtended,[12] was about 31′, answerable to the Suns Diameter; but the angle, which its length subtended, was more than five such diameters, namely 2 deg. 49′.

Having made these observations, I first computed from them the refractive power of that glass, and found it measured by the *ratio* of the sines, 20 to 31. And then, by that *ratio*, I computed the Refractions of two Rays flowing from opposite parts of the Sun's *discus,* so as to differ 31′ in their obliquity of Incidence, and found, that the emergent Rays should have comprehended an angle of about 31′, as they did, before they were incident.

The first of these two paragraphs is straightforward measurement; only its last sentences touch on a (simple) matter of theory. Newton tells us the angles of the prism, the size of the hole, and so on. Some of these things are relatively easy to measure with the precision he gives (like the angles of the prism, which do not vary); others are rather harder, either because they require one to make a decision about what is important and what is not (the length of the spectrum, for instance, depends on distinguishing stray light from the light that produces the spectrum proper) or demand special care in setting up the experiment and great precision and speed in executing it (the angles at which the beam is incident and emergent can be measured exactly only if the prism is fixed rigidly in place, and it requires some speed because the sun's motion through the sky changes the angle of incidence by 1 degree every 4 minutes, or 1 minute of arc every 4 seconds). Newton probably needed the help of an assistant, and it seems likely that he reached the precision of his measurements of the angles of incidence and refraction from theoretical considerations as well as from instruments like compass and protractor. The measurements tell us in addition that the prism was in the preferred orientation, the position of minimum deviation.

Newton's measurements show that the breadth of the spectrum stays in proportion to the size of the sun; it corresponds to about ½ degree of arc. The breadth is the same as if you simply let the light of the sun pass unrefracted to a screen at a distance of 22 feet; thus the breadth does not require any special explanation, it follows the principles of the linear propagation of light. But the spectrum's length is five times greater than the breadth of the sun. It seems that refraction is responsible for stretching out the image in one direction only; since he describes the boundaries as straight (except for the semicircular ends), it appears that refraction somehow spreads the light in a direction perpendicular to the refracting edge of the prism. But how does this come about?

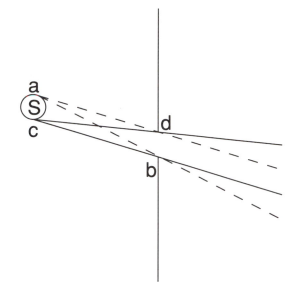

Fig. 12. In passing through the aperture db the rays coming from one extreme of the sun, a, are contained in the area between ad and ab, while those from the opposite extreme, c, are contained between cd and cb; and all other rays from intermediate points of the sun are therefore contained between the extreme rays ab and cd once they have passed through the aperture. If a prism can be oriented so that the extreme rays ab and cd minimally diverge from their original path, then the divergence or spreading of the spectrum as a whole will be minimized.

Three casual attempts at explanation failed. (1) The scattering hypothesis assumed that there was an initial refraction, the direction of which was determined according to the sine law, whereupon irregularities in the glass would cause some part of the light to be randomly dispersed, causing the image to spread. But if this effect is random, it should make the image larger in *all* directions. Thus the spectrum cannot be the result of *random local disturbances* in an otherwise regular beam. (2) The shadow or bounding hypothesis assumed that some disturbance at the edges of the beam caused the spreading, but putting the prism on the other side of the hole discounted this. Thus the spectrum is not the result of *edge disturbances* in an otherwise regular beam. (3) The glass-thickness hypothesis argued that a differentiation in the beam was induced by the different thickness of glass through which the opposite sides of the beam passed.[13*] This, too, must be rejected, but the hypothesis nevertheless foreshadows something of importance: that one must consider what happens to the different parts of the beam of light. Newton's attention in fact quickly turns to the parts of light, to *rays*. He pays attention to two rays in particular, the extreme rays whose paths are most oblique to one another (i.e., have the greatest angle between them; see Fig. 12).

It is precisely with respect to these rays that the position of minimum deviation is important. The sine law predicts that the extreme rays will be refracted so as to produce some elongation at all positions except those near minimum deviation. To really know these things one must of course prove them mathematically—something that Newton had expressly done in his optical lectures of 1670–1672 and was certainly capable of doing before then. Such mathematical proofs and calculations allow him to specify exactly how much of the elongation is unexplained by existing theories and hypotheses. From this point forward Newton is hot on the trail of something that can be determined in precise, quantitative terms. Since the mathematics of the existing theories shows there is still a problem to be solved, he turns back to experimentation to see what further light can be shed on the phenomenon.

But because this computation was founded on the Hypothesis of the proportionality of the *sines* of Incidence, and Refraction, which though by my own Experience I could not imagine to be so erroneous, as to make that Angle but 31′, which in reality was 2 deg. 49′; yet my curiosity caused me again to take my Prisme. And having placed it at my window, as before, I observed, that by turning it a little about its *axis* to and fro, so as to vary its obliquity to the light, more than an angle of 4 or 5 degrees, the Colours were not thereby sensibly translated from their place on the wall, and consequently by that variation of Incidence, the quantity of Refraction was not sensibly varied. By this Experiment therefore, as well as by the former computation, it was evident, that the difference of the Incidence of Rays, flowing from divers parts of the Sun, could not make them after decussation [the point where nonparallel rays cross] diverge at a sensibly greater angle, than that at which they before converged; which being, at most, but about 31 or 32 minutes, there still remained some other cause to be found out, from whence it could be 2 degr. 49′.

The circumstances of the experiment, translated into the terms of the mathematics of rays, lead to the hypothetical conclusion that a ½-degree difference in incidence of the beam's extreme rays would have to be sufficient to explain the fivefold elongation observed. By taking up the prism again to see what happens when it is rotated around the axis parallel to the refracting edge, Newton shows that, at least near the position of minimum deviation, even a 4- to 5-degree difference does not much affect the overall refraction of the image. Much less, then, can the ½-degree difference be the cause of elongation. This leads Newton to the suspicion that there is something wrong with the sine law, that at the very least in its accepted form it leaves something out of account. For a moment there rises the specter of having to give up one of the major achievements of seventeenth-century optics, the sine law governing refraction.

Is it possible, however, that the sine law can be preserved by an appropriate *revision?* Suppose the sine law is right about what happens *in* the prism but that something more happens after the light emerges. This is the sort of question that Newton entertains next.

Then I began to suspect, whether the Rays, after their trajection [passage] through the Prisme, did not move in curve lines, and according to their more or less curvity tend to divers parts of the wall. And it increased my suspition, when I remembered that I had often seen a Tennis ball, struck with an oblique Racket, describe such a curve line. For, a circular as well as a progressive motion being communicated to it by that stroak [stroke], its parts on that side, where the motions conspire, must press and beat the contiguous Air more violently than on the other, and there excite a reluctancy and reaction of the Air proportionably greater. And for the same reason, if the Rays of light should possibly be globular bodies, and by their oblique passage out of one medium into another acquire a circulating motion, they ought to feel the greater resistance from the ambient Aether, on that side, where the motions conspire, and thence be continually bowed to the other [Fig. 13]. But notwithstanding this plausible ground of suspition, when I came to examine it, I could observe no such curvity in them. And besides (which was enough for my purpose) I observed, that the difference 'twixt the length of the Image, and diameter of the hole, through which the light was transmitted, was proportionable to their distance.

In golf, tennis, baseball, and the like you can make a ball curve more or less sharply by applying spin. This suggests the hypothesis that little "balls" of light are set spinning by refraction and start to curve once they leave the prism. Now, lacking other evidence, this is quite a speculative hypothesis, which seems to commit one to a number of presuppositions

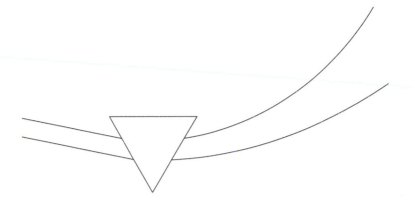

Fig. 13. A semi-Cartesian conjecture about the cause of the spectrum's elongation: some difference in the spin of tiny light globules after refraction might
produce different degrees of curvature in the paths of the rays.

(for instance that there are globe-like particles of light, and that some
hypothetical "ambient ether"—a thin but resilient air-like substance that
exists even where all air has been removed—plays a role in setting up a
difference of pressure that causes the curvature of the lightballs). This
hypothesis, like the earlier ones, is not entirely Newton's invention. It is
a more or less free variation on Descartes's explanation of color in the
Meteorology, which suggested that the little globules of light-transmitting matter produce different colors according to their rate and direction
of spin.[14] Fortunately there is no need to pursue this adaptation of
Descartes further, for experiments do not display any curvature in the
path at all. Presumably Newton took a screen (e.g., a sheet of paper) and
put it at increasing distances from the prism, and determined that the
difference between the length and breadth of the spectrum (length minus
width) constantly and steadily increased from the point where the light
left the prism.

What should one do next? Entertain more hypotheses, including one's
own? In the version of the letter that Newton sent to the Royal Society
there was a relevant passage eight paragraphs further on that was sub-
sequently dropped from the printed version. It went like this:

A naturalist would scearce [scarcely] expect to see ye [the] science of those
[colors] become mathematicall, & yet I dare affirm that there is as much certainty
in it as in any other part of Opticks. For what I shall tell concerning them is not an

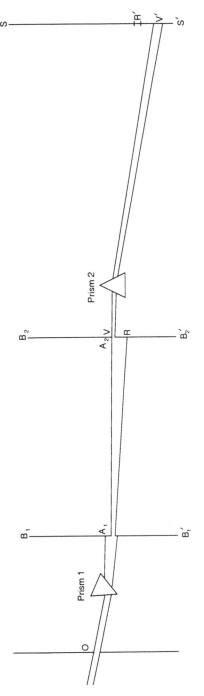

Fig. 14. The experimentum crucis. Boards B_1B_1' and B_2B_2' are fixed in place, as is prism 2; Prism 1 can be rotated to allow different parts of the spectrum VR to fall on the aperture A_2 in board B_2B_2'. When violet light passes through A_2, it is refracted toward region V' on screen SS', whereas red is less refracted, toward R'. Since the two boards and the second prism are fixed rigidly in place, the angle of incidence on Prism 2 does not vary; any difference in the position of the image on the screen is therefore due to a difference in the refraction of the beam in prism 2.

Hypothesis but most rigid consequence, not conjectured by barely inferring 'tis thus because not otherwise or because it satisfies all phaenomena (the Philosophers universall Topick,) but evinced by ye mediation of experiments concluding directly & wthout any suspicion of doubt.[15]

The point Newton makes in this passage is that the kind of science he is proposing is different from the *hypothetical* method used by most of his contemporaries. The hypothetical method works by proposing an unverified and perhaps unverifiable mechanism underlying appearances: for example, if light is supposed to consist of tiny, spinning globes, we can explain the elongation of the spectrum by saying they behave like a tennis ball given back- or topspin, or if we imagine that at the refracting surface a ray is given a sharp acceleration, as though it were a tennis ball struck by a racket, we can account for the change of direction in refraction. Newton found this method to be profoundly unsatisfying, since there is no end to possible hypotheses and since they encourage the researcher to undertake uncontrollable flights of imagination. Throughout his career Newton expressed a distrust of hypotheses, and although he did not always strictly observe the dictum he pronounced some fifteen years later in the *Principia* (*Hypotheses non fingo:* I do not frame or make up hypotheses), it reflected an enduring attitude. He believed that it was possible for natural philosophy to provide greater certainty than hypotheses. The next experiment was therefore intended to show that one could arrive at some real truths about light, even if they did not answer every question. It was an experiment that he called crucial.[16]

The gradual removal of these suspitions, at length led me to the *Experimentum Crucis,* which was this [Fig. 14]: I took two boards, and placed one of them close behind the Prisme at the window, so that the light might pass through a small hole, made In it for tho purpose, and fall on the other board, which I placed at about 12 feet distance, having first made a small hole in it also, for some of that Incident light to pass through. Then I placed another Prisme behind this second board, so that the light, trajected [passed] through both the boards, might pass through that also, and be again refracted before it arrived at the wall. This done, I took the first Prisme in my hand, and turned it to and fro slowly about its *Axis,* so much as to make the several parts of the Image, cast on the second board, successively pass through the hole in it, that I might observe to what places on the wall the second Prisme would refract them. And I saw by the variation of those places, that the light, tending to that end of the Image, towards which the refraction of the first Prisme was made, did in the second Prisme suffer a Refraction considerably greater then the light tending to the other end. And so the true cause of the length of that Image was detected to be no other, then that *Light* consists of *Rays differently refrangible,* which, without any respect to a difference in their incidence, were, according to their degrees of refrangibility, transmitted towards divers parts of the wall.

All the previous experiments served to refute hypotheses; this one, to the contrary, is intended to allow one to recognize a true and invariant property of light. Newton does not present the crucial experiment to confirm a hypothesis: no hypothesis has been enunciated. The crucial experiment is not deduced by logic from premises. It is not an inductive generalization, though one might argue that Newton is trying to develop a new variety of induction. Induction is ordinarily a process in which one examines a large number of cases of one kind or, where possible, all possible cases in order to arrive at a true generalization. Here there is just a single experiment to justify the conclusion. But in Newton's eyes this is a very special experiment, an experimentum crucis—an experiment made at a crossing point—which, by producing a striking phenomenon, is supposed to prove a theory, a way of seeing things truly.

In the second part of the *New Organon* Francis Bacon (1561–1626) explained how experience could be employed to arrive at a true assessment of the forms and natures of things, and he provided a classification of different kinds of experiences and instances according to how they contributed to understanding. One of these was the *instantia crucis,* the instance of the crossing. The name plays on the image of the intersection of two or more roads: some experiences or instances are designed to place one at the crux of several possibilities, and the outcome can definitively rule out one or more of them as unviable and perhaps even point to a single one as the correct way.[17]

What is the crossing at which Newton stands? A full answer can be inferred from both his earlier optical lectures and the subsequent correspondence debating the status of his new theory. Newton saw the field dominated by a conception of light according to which colors and the elongation of the refracted image were produced by some modification or differentiation caused by the encounter of ordinary, simple light with matter. Thus as in the hypothesis attributed to Dominis, a longer path through the prism (an encounter with a greater quantity of matter) modified the light so that it spread and changed in color, or (following Descartes) the different reactions at the edge between light and darkness led to the light's coloration and dispersion.

Within this modificationist conception the basic entity was simple white light, which was capable of taking on different qualities. The modificationist could reason in the following way: Light as light is of one kind; it is homogeneous. This homogeneous light (let us call it "illuminant") can take on different qualitative appearances depending on how it is affected by its encounter with matter, but it still remains essentially illuminant, and so its essence and its fundamental behavior remain the same. When refracted, illuminant is turned out of its original path according to the sine law. The spreading that Newton has emphasized, continues the modificationist, is not caused by the refraction of the illuminant as such; rather, some as yet not completely understood property or characteristic of the illuminant is either produced or affected by the circumstances of refraction (for example, it is mixed with darkness, or its particles start spinning).

What the crucial experiment does is isolate a specific phenomenon that helps display a fundamental behavior of light in its encounter with the prism. The refraction by the second prism shows that the portion of the light that was refracted to the greatest extent by the first prism (the light in the violet part of the spectrum—but note that Newton does not even mention colors here) is also refracted to the greatest extent by the second. In addition, the light refracted a second time does not produce an elongated image (a fact that Newton does not state here but only in later correspondence). Thus refracted light behaves differently from unrefracted light, so that one cannot argue that all light as illuminant behaves essentially the same. The conclusion: light is not a uniform thing that uniformly obeys the law of sines but rather is composed of diverse kinds of light, each of which obeys a sine law, a sine proportion, that is slightly different for each kind of light. Since there are no discontinuities in the spectrum, for each value of the sine proportion between the extremes there corresponds a kind of light refracted according to that proportion. Since there is no limit to possible intermediate values between any two sine proportions, there may be no limit to the number of intermediate kinds of light.

It is clear, of course, that the experiments described in the letter are sufficient only to suggest rather than to substantiate all these points. A more thorough and more mathematical approach already existed in the form of the lectures Newton produced between 1670 and 1672 at Cambridge University. In the *Opticks* of 1704 Newton repeated and refined the experimental substantiation. For one thing, experiments must clearly show that the images produced by the second prism are not elongated, that is, are very nearly circular (recall that Newton had proved mathematically that the image should be circular for light refracted according to a single sine proportion in the position of minimum deviation). What the crucial experiment shows, though this also is not perfectly explicit, is that as one rotates the first prism and successively casts red, then orange, then yellow, then green, then blue, then indigo, then violet on the second aperture, the nearly circular image moves across the wall or screen from the minimal refraction of red to the maximal refraction of violet.

It is important to emphasize that Newton does not mention colors at all in the account of the experimentum crucis; in fact, he is very sparing in the use of color names in the first half of the letter. It is only in the second half that he takes up color itself. In the subsequent debate about the theory Newton claimed that he had proved the differential refrangibility of light without reference to color, and thus that it was true whatever the relation between colors and kinds of light turns out to be. Yet probably the best one can hope for here is a *relative* separation of the issues. The light of the least-refracted parts of the first spectrum is least refracted in the second refraction, it is true, but if the color of the light changed significantly it would be hard to argue that one had isolated rays with an unvarying property or a unique identity.

Before moving on to the topic of colors, however, Newton makes perfectly clear what the connection is between his work with lenses, his theory of differential refrangibility in prismatic refraction, and his invention of a reflecting telescope:

When I understood this, I left off my aforesaid Glass works; for I saw, that the perfection of Telescopes was hitherto limited, not so much for want of glasses truly figured according to the prescriptions of Optick Authors, (which all men have hitherto imagined,) as because that Light it self is a *Heterogeneous mixture of differently refrangible Rays.* So that, were a glass so exactly figured, as to collect any one sort of Rays into one point, it could not collect those also into the same point, which having the same Incidence upon the same Medium are apt to suffer a different refraction. Nay, I wondered, that seeing the difference of refrangibility was so great, as I found it, Telescopes should arrive to that perfection they are now at. For, measuring the refractions in one of my Prismes, I found, that

Fig. 15

supposing the common *sine* of Incidence upon one of its planes was 44 parts, the *sine* of refraction of the utmost Rays on the red end of the Colours, made out of the glass into the Air, would be 68 parts, and the *sine* of refraction of the utmost rays on the other end, 69 parts: So that the difference is about a 24*th* or 25*th* part of the whole refraction.

Newton expresses the ratio of the sines of incidence and refraction from glass to air as a ratio between whole numbers rather than as a decimal fraction, a common practice in his day. (For us it is more common to say that the sine ratio for the least refrangible rays is 1.545 rather than 68 to 44). If for any given refraction from the glass into air we represent the sine of incidence by a line segment, we can divide it into 44 parts (YP in Fig. 15) and then create two longer segments, one 68 units long (YR), the other 69 units (YV). The proportion of the 68-unit segment to the 44-unit segment then stands for the sine relation of incidence to refraction for the least-refrangible red-producing rays, whereas the 69-unit segment stands to the 44-unit segment in the same proportion as the sine of incidence to the sine of refraction for the most-refractible violet-producing rays. Refraction thus adds to the sine from the angle of incidence between 24 and 25 units, depending on the color, and the disparity between the sine of the extreme red and extreme violet rays (one unit) is between 1/24 and 1/25 of the total refraction of 24–25 units.

The preceding paragraph then concludes:

And consequently, the object-glass of any Telescope cannot collect all the rays, which come from one point of an object so as to make them convene at its *focus* in less room then in a circular space, whose diameter is the 50*th* part of the Diameter of its Aperture; which is an irregularity, some hundreds of times greater, then a circularly figured *Lens,* of so small a section as the Object glasses of long Telescopes are, would cause by the unfitness of its figure, were Light *uniform.*

The question Newton is implicitly asking is this: how much more difficult is it to bring all the rays to a focus if, in addition to the ordinary aberration due to the spherical curvature of a lens (spherical aberration), one adds the differential refrangibility according to color (chromatic aberration)? According to a formula I assert but do not prove here, if the common sine of incidence is represented by the number I, the sine of refraction of the least refrangible rays by R, and that of the most refrangible rays by S, then the diameter of the smallest possible circular area on which the light can be focused is $(S - R)/(S + R - 2I)$ times the size of the aperture (that is, of the portion of the lens that actually transmits the refracted beam of light, which in most cases is simply the diameter of the entire lens). Perfect focus would bring all the rays to a mathematical point. With $I = 44$, $R = 68$, and $S = 69$, we get $(69 - 68)/(69 + 68 - 2 \times 44)$ = 1/49: about $\frac{1}{50}$ of the diameter of the aperture, as Newton says.

This made me take *Reflections* into consideration, and finding them regular, so that the Angle of Reflection of all sorts of Rays was equal to their Angle of Incidence; I understood, that by their mediation Optick instruments might be brought to any degree of perfection imaginable, provided a *Reflecting* substance could be found, which would polish as finely as Glass, and *reflect* as much light, as glass *transmits,* and the art of communicating to it a *Parabolick* figure be also attained. But there seemed very great difficulties, and I have almost thought them insuperable, when I further considered, that every irregularity in a reflecting superficies makes the rays stray 5 or 6 times more out of their due course, than the like irregularities in a refracting one: So that a much greater curiosity [precision] would be here requisite, than in figuring glasses for Refraction.

Amidst these thoughts I was forced from *Cambridge* by the Intervening Plague, and it was more then two years, before I proceeded further. But then having thought on a tender way of polishing, proper for metall, whereby, as I imagined, the figure also would be corrected to the last; I began to try, what might be effected in this kind, and by degrees so far perfected an Instrument (in the essential parts of it like that I sent to *London,*) by which I could discern Jupiters 4 Concomitants [moons], and shewed them divers times to two others of my acquaintance. I could also discern the Moon-like phase of *Venus,* but not very distinctly, nor without some niceness in disposing the Instrument.

From that time I was interrupted till this last Autumn, when I made the other. And as that was sensibly better then the first (especially for Day-Objects,) so I doubt not, but they will be still brought to a much greater perfection by their endeavours, who, as you inform me, are taking care about it at *London.*

I have sometimes thought to make a *Microscope,* which in like manner should have, instead of an Object-glass, a Reflecting piece of metall. And this I hope they will also take into consideration. For those Instruments seem as capable of improvement as *Telescopes,* and perhaps more, because but one reflective piece of metall is requisite in them. . . .

The connection with his work on telescopes is explicit here: although optical workers had hitherto tried to minimize the distortion of lenses by trying to grind shapes other than spherical, the distortion due to color, chromatic aberration, is far more important than the shape of the lens in preventing the perfection of telescopes. But *reflection* does not differ according to the different kinds of rays, and so a reflecting or *catoptric* telescope avoids the major cause of distortion and should produce a superior instrument, if the mirror can be given a parabolic shape (as in automobile headlights). What remains to achieve is a certain technical facility and appropriate materials for producing such a telescope; the problem in theory is resolved.

Thus far Newton's letter has presented one major theoretical result, the compositeness of light, which is separable according to degree of refraction, and the resulting practical consequence, that refracting telescopes face a nearly insuperable obstacle to their improvement and so ought to be replaced by reflectors. But the most surprising thing of all is that there is a theoretical bonus, a further development of the theory of differential refrangibility: a new theory of color virtually unprecedented in the annals of natural research. So after summarizing the conclusion of the first part of the letter, Newton presents his theory of color, in thirteen propositions.

But to return from this digression, I told you, that Light is not similar, or homogeneal, but consists of *difform* Rays, some of which are more refrangible than others: So that of those, which are alike incident on the same medium, some shall be more refracted than others, and that not by any virtue of the glass, or other external cause, but from a predisposition, which every particular Ray hath to suffer a particular degree of Refraction.

I shall now proceed to acquaint you with another more notable difformity in its Rays, wherein the *Origin of Colours* is unfolded: Concerning which I shall lay down the *Doctrine* first, and then, for its examination, give you an instance or two of the *Experiments,* as a specimen of the rest.

The Doctrine you will find comprehended and illustrated in the following propositions.

1. As the Rays of light differ in degrees of Refrangibility, so they also differ in their disposition to exhibit this or that particular colour. Colours are not *Qualifica-*

tions of Light, derived from Refractions, or Reflections of natural Bodies (as 'tis generally believed,) but *Original* and *connate properties,* which in divers Rays are divers. Some Rays are disposed to exhibit a red colour and no other; some a yellow and no other, some a green and no other, and so of the rest. Nor are there only Rays proper and particular to the more eminent colours, but even to all their intermediate gradations.

2. To the same degree of Refrangibility ever belongs the same colour, and to the same colour ever belongs the same degree of Refrangibility. The *least Refrangible* Rays are all disposed to exhibit a *Red* colour, and contrarily those Rays, which are disposed to exhibit a *Red* colour, are all the least refrangible: So the *most refrangible* Rays are all disposed to exhibit a deep *Violet Colour,* and contrarily those which are apt to exhibit such a violet colour, are all the most Refrangible. And so to all the intermediate colours in a continued series belong intermediate degrees of refrangibility. And this Analogy 'twixt colours, and refrangibility, is very precise and strict; the Rays always either exactly agreeing in both, or proportionally disagreeing in both.

3. The species[18] of colour, and degree of Refrangibility proper to any particular sort of Rays, is not mutable by Refraction, nor by Reflection from natural bodies, nor by any other cause, that I could yet observe. When any one sort of Rays hath been well parted from those of other kinds, it hath afterwards obstinately retained its colour, notwithstanding my utmost endeavours to change it. I have refracted it with Prismes, and reflected it with Bodies, which in Day-light were of other colours; I have intercepted it with the coloured film of Air interceding [between] two compressed plates of glass; transmitted it through coloured Mediums, and through Mediums irradiated with other sorts of Rays, and diversly terminated it; and yet could never produce any new colour out of it. It would by contracting or dilating become more brisk, or faint, and by the loss of many Rays, in some cases very obscure and dark; but I could never see it changed *in specie* [kind].

4. Yet seeming transmutations of Colours may be made, where there is any mixture of divers sorts of Rays. For in such mixtures, the component colours appear not, but, by their mutual allaying each other, constitute a midling colour. And therefore, if by refraction, or any other of the aforesaid causes, the difform Rays, latent in such a mixture, be separated, there shall emerge colours different from the colour of the composition. Which colours are not New generated, but only made Apparent by being parted; for if they be again intirely mix't and blended together, they will again compose that colour, which they did before separation. And for the same reason, Transmutations made by the convening of divers colours are not real; for when the difform Rays are again severed, they will exhibit

the very same colours, which they did before they entered the composition; as you see, *Blew* and *Yellow* powders, when finely mixed, appear to the naked eye *Green,* and yet the Colours of the Component corpuscles are not thereby really transmuted, but only blended. For, when viewed with a good Microscope, they still appear *Blew* and *Yellow* interspersedly.[19]

5. There are therefore two sorts of Colours. The one original and simple, the other compounded of these. The Original or primary colours are, *Red, Yellow, Green, Blew,* and a *Violet-purple,* together with Orange, Indico [indigo], and an indefinite variety of Intermediate gradations.

6. The same colours in *Specie* with these Primary ones may be also produced by composition: For, a mixture of *Yellow* and *Blew* makes *Green;* of *Red* and *Yellow* makes *Orange;* of *Orange* and *Yellowish green* makes *yellow.* And in general, if any two Colours be mixed, which in the series of those, generated by the Prisme, are not too far distant one from another, they by their mutual alloy compound that colour, which in the said series appeareth in the mid-way between them. But those, which are situated at too great a distance, do not so. *Orange* and *Indico* produce not the intermediate Green, nor Scarlet and Green the intermediate yellow.

7. But the most surprising, and wonderful composition was that of *Whiteness.* There is no one sort of Rays which alone can exhibit this. 'Tis ever compounded, and to its composition are requisite all the aforesaid primary Colours, mixed in a due proportion. I have often with Admiration beheld, that all the Colours of the Prisme being made to converge, and thereby to be again mixed as they were in the light before it was Incident upon the Prisme, reproduced light, intirely and perfectly white, and not at all sensibly differing from a *direct* Light of the Sun, unless when the glasses, I used, were not sufficiently clear; for then they would a little incline it to *their* colour.

8. Hence therefore it comes to pass, that *Whiteness* is the usual colour of *Light;* for, Light is a confused aggregate of Rays indued [endowed] with all sorts of Colors, as they are promiscuously darted from the various parts of luminous bodies. And of such a confused aggregate, as I said, is generated Whiteness, if there be a due proportion of the Ingrodiontc; but if any one predominate, the Light must incline to that colour; as it happens in the Blew flame of Brimstone [sulfur]; the yellow flame of a Candle; and the various colours of the Fixed stars.

The "doctrine of colors" presented in the first eight propositions is the historical basis of the modern physical theory of color. According to Newton's presentation, light is not simple but complex, as proved by different refrangibility; the difference in refrangibility is grounded in some (up to this point unidentified) characteristic that differentiates rays, and it corresponds very precisely to differences in colors. This differentiation is continuous rather than discrete: there are rays corresponding not just to the principal species of color (blue, red, green, and so forth), but to all intermediate colors as well; thus the number of colors is infinite, or at least indefinitely large. (Newton calls the colors produced by a single kind of ray *original* or *primary*—but note that the latter term means something different from the modern "primary colors.")[20]

The experimentum crucis showed that rays retain their degree of refrangibility through multiple refractions; since the colors also remain the same, this (and other experiments) shows that the color-producing property of colors is unchanged. There is thus a one-to-one correspondence between refrangibility and color. A complication is that mixing different kinds of rays results in an intermediate, composite color, which in many cases is virtually indistinguishable from a simple spectral color. In some cases combination produces a new color not seen in the spectrum (like the purple derived from red and violet, and, of course, white).[21*] But in every case of such composite colors, Newton asserts, the components can be separated again, so that it would seem to be incorrect to say that mixing rays produces a "transmutation." Just as in a powder consisting of very fine, differently colored particles the individual colors are present, although not distinguishable to ordinary vision, so does a beam of light consist of a mixture of different color-producing rays.

Color theory before Newton had treated the colors produced by refracted light in a special category: they were called *emphatic* or *apparent colors* and were thought to be of secondary importance in comparison with the real colors of bodies. Newton is inverting this hierarchy. He is arguing that the so-called apparent colors are fundamental, whereas the supposedly real colors turn out to result from the disposition of bodies to reflect this or that kind of ray more or less abundantly. In homogeneal blue light a red apple will look blue, so it no longer makes sense to talk about the apple's real color being red. Accordingly, in the next five theses Newton explains not just emphatic colors like those of the prismatic spectrum and the rainbow but also all the apparently more ordinary phenomena of color in terms of the separation and mixture of color-producing rays.

9. These things considered, the *manner,* how colours are produced by the Prisme, is evident. For, of the Rays, constituting the incident light, since those

which differ in Colour proportionally differ in Refrangibility, *they* by their unequall refractions must be severed and dispersed into an oblong form in an orderly succession from the least refracted Scarlet to the most refracted Violet. And for the same reason it is, that objects, when looked upon through a Prisme, appear coloured. For, the difform Rays, by their unequal Refractions, are made to diverge towards several parts of the *Retina,* and there express the Images of things coloured, as in the former case they did the Suns Image upon a wall. And by this inequality of refractions they become not only coloured, but also very confused and indistinct.

10. Why the Colours of the *Rainbow* appear in falling drops of Rain, is also from hence evident. For, those drops, which refract the Rays, disposed to appear purple, in greatest quantity to the Spectators eye, refract the Rays of other sorts so much less, as to make them pass beside it; and such are the drops on the inside of the *Primary* Bow, and on the outside of the *Secondary* or Exteriour one.[22] So those drops, which refract in greatest plenty the Rays, apt to appear red, toward the Spectators eye, refract those of other sorts so much more, as to make them pass beside it; and such are the drops on the exteriour part of the *Primary,* and interiour part of the *Secondary* Bow.

11. The odd Phaenomena of an infusion of *Lignum Nephriticum,*[23] *Leaf gold, Fragments of coloured glass,* and some other transparently coloured bodies, appearing in one position of one colour, and of another in another, are on these grounds no longer riddles. For, those are substances apt to reflect one sort of light and transmit another; as may be seen in a dark room, by illuminating them with similar or uncompounded light. For, then they appear of that colour only, with which they are illuminated, but yet in one position more vivid and luminous than in another, accordingly as they are disposed more or less to reflect or transmit the incident colour.

12. From hence also is manifest the reason of an unexpected Experiment, which Mr. *Hook* [Robert Hooke] somewhere in his *Micrography* [*Micrographia*] relates to have made with two wedg-like transparent vessels, fill'd the one with a red, the other with a blew liquor: namely, that though they were severally transparent enough, yet both together became opake; For, if one transmitted only red, and the other only blew, no rays could pass through both.

13. I might add more instances of this nature, but I shall conclude with this general one, that the Colours of all natural Bodies have no other origin than this, that they are variously qualified to reflect one sort of light in greater plenty then another. And this I have experimented in a dark Room by illuminating those bodies with uncompounded light of divers colours. For by that means any body may be made to appear of any colour. They have there no appropriate colour, but

ever appear of the colour of the light cast upon them, but yet with this difference, that they are most brisk and vivid in the light of their own day-light-colour. *Minium* [red lead] appeareth there of any colour indifferently, with which 'tis illustrated [illuminated], but yet most luminous in red, and so *Bise* [or bice, a deep blue pigment obtained from smalt] appeareth indifferently of any colour with which 'tis illustrated, but yet most luminous in blew. And therefore *Minium* reflecteth Rays of any colour, but most copiously those indued with red; and consequently when illustrated with day-light, that is, with all sorts of Rays promiscuously blended, those qualified with red shall abound most in the reflected light, and by their prevalence cause it to appear of that colour. And for the same reason *Bise,* reflecting blew most copiously, shall appear blew by the excess of those Rays in its reflected light; and the like of other bodies. And that this is the intire and adequate cause of their colours, is manifest, because they have no power to change or alter the colours of any sort of Rays incident apart, but put on all colours indifferently, with which they are inlightned.

Thus it becomes clear that Newton is proposing no less than a comprehensive theory of light and colors and, implicitly, a theory of the physical basis of seeing as well. The appearance of all things revealed by light is determined by the kind and quantity of rays that those things reflect (or let pass) to the eye. Whatever appears blue does so because in the light reflected from it there is a predominance of "blue rays," and similarly for the appearances of other colors. And since any object, if it is illuminated with homogeneal or monochromatic light, will reflect just that one kind of light and thus appear in the corresponding color, objects cannot be said to *have* color in the strict sense. The theory provides a pattern of explanation; in terms made current by recent philosophy of science, Newton's theory of light and colors offers a *paradigm* that takes possession of the field of optics and provides the fundamental terms and approaches in which that field will be understood.

These things being so, it can be no longer disputed, whether there be colours in the dark, nor whether they be the qualities of the objects we see, no nor perhaps, whether Light be a Body. For, since Colours are the *qualities* of Light, having its Rays for their intire and immediate subject, how can we think those Rays *qualities* also, unless one quality may be the subject of and sustain another; which in effect is to call it *Substance.* We should not know Bodies for substances, were it not for their sensible qualities, and the Principal of those being now found due to something else, we have as good reason to believe that to be a Substance also.

Besides, whoever thought any quality to be a *heterogenous* aggregate, such as Light is discovered to be. But, to determine more absolutely, what Light is, after what manner refracted, and by what modes or actions it produceth in our minds the Phantasms [images in consciousness] of Colours, is not so easie. And I shall not mingle conjectures with certainties.

These two paragraphs constitute a more general, philosophical reflection on what Newton has established. Newton has settled certain traditional questions. Are there colors in the dark? No, because the color-producing quality is inherent in rays of light. Do the colors we see inhere in the things we see; are apples red and bananas yellow? No; these things may have a tendency to reflect more red-producing rays or yellow-producing rays, but the colors themselves are not in the objects, they are a characteristic of the rays. Newton even believes (though he is hesitant to assert it positively) that he has established that light must be some kind of body. The argument is philosophical or ontological. The traditional philosophy based on Aristotelian concepts divides being into two kinds: a thing is either a substance, which is (at least relatively) independent of other things, or it is an accident (or characteristic or, more loosely, quality), which can exist only if it belongs to or inheres in a substance. If color is a quality, then it must inhere in a substance; Newton has shown that it does not inhere in the objects that appear colored but rather in the rays that produce it; so if color inheres in something, it makes sense to say it is in the rays. But then rays must be substances, because if they were not they would be qualities instead, and so colors would be qualities of qualities, something that traditional metaphysics forbids.

Newton now includes a final experiment (see Fig. 16). In the darkened room he refracts light with prism ABC, intercepts it with a focusing lens MN, and shows with a movable screen HI that as the light approaches the point of focus Q, the spectral colors (violet and blue nearer H, red and yellow nearer I) begin to disappear until at Q the image turns perfectly white again; then, as the rays start diverging once more past Q, the colors reappear in reverse order (violet and blue nearer I, red and yellow nearer H). The experiment is meant to establish a crucial principle of Newton's theory of color. If the theory is right, it should be possible to separate rays from one another and then recombine them without any effect whatsoever on their color-producing characteristics. If in any way they should bear marks of their having been previously refracted, then some version of the modificationist theory of color would be true. But since,

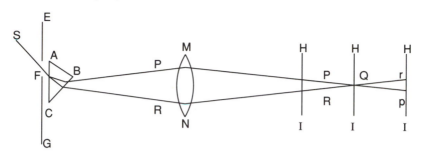

Fig. 16

after they have been separated, the rays can be combined again into a white color indistinguishable from that made by the light of the sun, Newton's main contention is verified.[24*]

Newton finishes the letter with admonitions that those trying to confirm his theory must take care in performing the experiments: in particular, they must make sure that the light rays they work with are well separated from one another to avoid spurious disconfirming results (e.g., if one has not separated the green rays thoroughly from the neighboring blue and yellow, further experiments might lead one to believe that a second refraction "decomposes" spectral green into blue and yellow). He concludes with an invitation to his readers:

> . . . This, I conceive, is enough for an Introduction to Experiments of this kind; which if any of the *R.*[oyal] *Society* shall be so curious as to prosecute, I should be very glad to be informed with what success: That, if any thing seem to be defective, or to thwart this relation, I may have an opportunity of giving further direction about it, or of acknowledging my errors, if I have committed any.

In the *Philosophical Transactions* of the Royal Society the editor concluded with a comment that this was a "Learned and very Ingenious Letter; which having been by that *Illustrious Company,* before whom it was read, with much applause committed to the consideration of some of their Fellows, well versed in this argument, the Reader may possibly in an other *Tract* be informed of some report given in upon this Discourse."[25] The reader was indeed treated to more about the subject, for some of the Fellows and other experts identified what they thought were defects and errors, and Newton was compelled to write several letters in response—as we shall see in the next chapter.

Chapter 4

From the Early Theory to the *Opticks*

The theory introduced to the public by the letter of 6 February 1672 seems unproblematic, especially to us, who have been taught to think in terms of rays and refrangibilities. Not everyone was satisfied, however, and among Newton's chief critics were people who are regarded as leading figures of the scientific revolution of the seventeenth century, such as Christiaan Huygens (1629–1695) and Robert Hooke (1635–1702). Although we shall not follow the history of the controversy in detail, it is useful to outline some of the major objections. These may be roughly divided into those concerning what we know (methodological and epistemological) and what we see (phenomenal and experimental).

First, Newton, by calling his new doctrine a *theory,* intended to distinguish it from hypotheses. As he said in a subsequent letter, he would rather have suppressed his discovery than let it be known if it were only a hypothesis, that is, only a possible or probable assumption.[1] So by calling it a theory he was claiming more for his doctrine than just probability: he intended it as a matter of fact, as established truth.

Virtually all the disputants admitted that his doctrine was ingenious and interesting, but they took exception to Newton's tone and claims of certainty, raised some experimental points that they thought needed resolution, and offered alternative hypotheses. One thing at issue was Newton's breach of "etiquette," in that the community of natural philosophers was accustomed to avoid claims to certainty and the kind of passion that was more characteristic of religious and political arguments.[2] Huygens, for instance, broke off correspondence after two letters because of the "heat with which Mr. Newton argued." Hooke and the French optician Ignace Pardies (1636–1673) also objected to Newton's certitude and pitted against his theory their own versions of impulse theories of light and modificationism with respect to color. Newton reacted with thinly veiled impatience to their persistence in arguing for modificationism, since his letter aimed at demolishing its fundamental claims. As for impulse theories, although he had not dealt with them directly, they would all seem to involve some variety of modificationism (white light would appear to be a simple pulse that is modified by refraction: but then why are not pulses corresponding to red, green, blue, and so forth further modified by additional refractions?); moreover, as he pointed out to Hooke, all known pulses (like water and sound waves) go around corners, whereas light, which forms distinct shadows, does not.

A number of experimental objections were raised by the opponents, though sometimes in the context of their own theories.[3] Among historians of science it was long claimed that these objections were caused mostly by misunderstandings or stubbornness, but in the past few decades it has become clearer that there was a point to at least some of them. For example, Anthony Lucas lined up colored strips at the bottom of a basin, filled it with water, and then, peering over the edge of the basin, raised and lowered his head so that the strips would appear and then disappear from view. He reasoned that, if light is refracted differently according to color, as one raises one's head the most refractible colors ought to come into view before the least refractible; but instead he saw all the colors appear at exactly the same moment. Newton's answer was that such conditions are not favorable for making exact discriminations (presumably the basin was white, and so the bright light from the white background tended to overwhelm the much lesser quantity of light from the colored strips), and he impatiently urged Lucas to forget about the variety of experiments that could be raised in objection and instead to concentrate on the crucial experiment.

Lucas had in fact described a whole series of experiments, some quite ingenious, that he thought raised difficulties for Newton's theory. Newton's response was testy, even angry. It should be mentioned in his defense that by the time Lucas's first letter arrived Newton had been engaged in disputes about his theory for more than four years and was weary of it. Some of the earliest objections had indeed been trivial or obtuse, and the three major opponents (Hooke, Pardies, and Huygens) all held to pulse theories of light, a fact that had led Newton to accuse them of excessive partiality for their own hypotheses, which he thought made them unwilling or unable to admit the truth of a contradicting theory. Moreover, it is fairly evident that, except perhaps for Huygens, Newton surpassed all his contemporaries in depth and breadth of knowledge about light and colors and about the theories that had been proposed to explain them. But it cannot be denied that Newton himself exhibited a partiality for his own theory that far exceeded that of his opponents, and he expected them to accept the truth and logic of the crucial experiment: that the examination of a single experiment, a crucial experiment, could by itself determine whether a theory was right.

What is wrong with such a notion? Can't a single experiment be decisive? Of course it can, if it is known to be reliably performed and if its result is *negative*. Philosophers of science are fond of saying that a single black swan shows the falsity of "All swans are white." But if the person doing the inventory of swans is not sufficiently careful (and even if he or she is), it may turn out that there was a misclassification (for example, it is a swan, but it is black because of an oil spill, or it is a black bird that resembles but is not a swan). The case of the black swan is not really an *experiment*, of course, just an *observation*.

One might suspect that determining whether the results of an experiment are reliable is even more complicated than simple observation. If you have been using a prism to try some of the phenomena discussed up to this point, you already realize that it takes practice to reproduce an experiment. If you have tried the crucial experiment, you recognize that it takes several tries (and some "fiddling about" with the equipment) to get a result like Newton's, but even then one will probably notice some things he does not mention—for example, that the colored images of the twice-refracted light are not quite as circular as one might like, and that they occasionally display unexpected fringes of other colors.

If you have taken a laboratory science such as chemistry, you know that it is sometimes very hard to get the results the lab book says you should (and not a few students "fudge" the data ever so slightly). What you must realize is that this "margin of uncertainty" in experiments is not an aberration that would be eliminated by greater expertise but is something characteristic of all experimentation. Over time you do become more proficient, and if you repeatedly try elementary experiments you more frequently come close to the theoretical ideal; still, sometimes for reasons you are able to trace, sometimes not, you get results that are wildly aberrant. In addition, as you become more expert you also come to expect greater accuracy. For the beginning student an error of $1/10$ gram in weighing a chemical might be acceptable, but for the professional $1/1000$ gram might not be good enough. Nevertheless, there will always be some margin of error that will set limits to the accuracy of an individual experiment or of a class of experiments.

The point is that it is not really a single experiment that refutes a hypothesis but at best a single class of experiments that need to be performed and varied until they consistently come out right and described so that they can be more or less reproduced by others; and in observing color phenomena, or taking measurements, or reading dials or computer printouts, one must make "reasonable" decisions about what is significant and what is not. If you see yellow light fringed very slightly with green and blue, you might decide (with Newton) that the experiment works pretty much as the theory says it should, only that the yellow light was not as pure as one thought; but if you refract light taken from the red end of a first spectrum and get a patch of red fringed with violet, you must begin wondering how violet, which belongs at the opposite extreme of the spectrum, got mixed in with the red. (This was in fact an objection raised in the 1680s by the distinguished French experimentalist and theoretician Edmé Mariotte.)

Can a single experiment or even type of experiment prove a hypothesis to be right? This is a form of what philosophers call the *problem of induction,* which asks how you can prove a hypothesis to be true when it is not possible to examine all instances in which the hypothesis should hold. The short answer is that you cannot eliminate every last residue of uncertainty in any but the most trivial and limited inductions (e.g., all the coins currently in my pocket are nickels—not a very interesting or useful truth). Relying on a single experiment or experiment type, therefore, would appear to be hopeless. Yet one of the oddities of history is that the crucial experiment came to be widely accepted in the sciences precisely under the influence of Newton's example in the letter.

Was Newton just naive about the power of a single experiment? Although that is a possibility, one usually does not get very far by thinking that men and women of intellectual prowess could not see things that are obvious to us. Perhaps we can make a case for Newton, not just with respect to the crucial experiment of the letter but also with respect to his other inductions from experiment, along these lines. The crucial experiment by its nature and name stands at a crossroad. The crossroad is the intersection of proposals that have been advanced to explain the thing in question, that is, the intersection of competing hypotheses and conceptions. Although it is impossible to enumerate all conceivable hypotheses concerning a phenomenon, it is certainly possible to list the various kinds that have been advanced and perhaps even to identify approaches that have not yet been adequately developed but offer at least a minimal promise. In a not entirely trivial sense this is obvious: for example, one could say that the cause of refraction is either physical or nonphysical (logically speaking this has to be true, but note that it does not exclude the possibility that there could be *both* physical and nonphysical causes). By showing that one of the two possibilities is not tenable, one proves that the other is true.

If we weaken the standard from covering all *logical* possibilities to covering instead all *plausible* or *concretely conceivable* possibilities—suppose there are five—then we could prove that a true theory must be of type 1 by excluding types 2 through 5. But each of those might be eliminated by a single experiment, so perhaps just four experiments would be sufficient to establish type 1 as true. The limitation here, of course, is that a theory *type* does not necessarily imply a fully articulated and exhaustive theory. This becomes clearer from considering Newton's situation. Newton was not claiming that he had solved every single problem connected with light and color, but rather that a certain class of possibilities—modificationist theories of light, of which impulse theories are a variety—was no longer defensible. The crucial experiment indeed raises very considerable problems for modificationist theories, because it shows that a first refraction modifies light very considerably, whereas a second modifies the (now colored) light only slightly or not at all.

It is characteristic of modification theories to contend that all light is essentially alike (there is a unique index of refraction for it, a large beam behaves in essentially the same way as a small one, and so on). The crucial experiment points away from this theoretical route because it shows that light is not always and everywhere the same, that there are different kinds of light, and that once one has such a kind in one's experimental possession, one cannot change it into another kind. At this point Newton's theory of knowledge and of the nature of things enters: because the "separated" light one obtains experimentally cannot be changed by further experiments, it must be *basic,* an *ultimate constituent* of light, and so it has the right to be called a substance. (An impulse, on the other hand, would properly speaking be some mode of its medium; e.g., a water wave is a mode of water and not an independent substance. See Appendix A.)

In the eighteenth century most physicists believed that Newton had established that light consisted of unchangeable small bodies or corpuscles. Although the 1672 letter suggests precisely this, and although the *Opticks* seems to allow no other possibility, Newton was careful not to assert this without qualification. Early in the nineteenth century physicists became convinced that corpuscularism was wrong, that instead light consists of waves, which came to be thought of as the oscillations of a medium called the ether. That is, light was conceived according to a version of the hated modificationism! Speaking from a historical perspective, we must conclude that Newton overrated the crucial experiment's value as proof. Yet doubtless there is still something in it that remains valid. The crucial experiment did show something new, that repeated refraction produces light with a property that cannot be further modified, at least not obviously modified, a fact that differentiates it from white sunlight. Insofar as this resultant light is a terminal point in the process of analyzing white light, it can be called *analytically fundamental*. It has a reliably consistent and identifiable property arrived at through a process of analysis that reaches an end, that does not manage to produce anything more basic.[4]

The nineteenth-century theorists did not see the triumph of wave theory as a complete overthrow of Newton's conception of light. Light could, of course, no longer be plausibly interpreted as a small particle, but the differential refrangibility of light and the correlation of refrangibility to color that had been so basic to Newton's conception were readily adaptable into the wave theory, in which the difference in the frequency and length of the oscillation of the wave were used to account for differences in refraction and in colors.[5] Newton's homogeneal light came to be reinterpreted as monochromatic light, that is, radiation of a single frequency or wavelength and representable by pure sine-wave forms. Yet although the nineteenth-century wave theory was more sophisticated than late-seventeenth-century impulse and wave theories, and although the correspondence of wavelength to refrangibility and color is analogous to Newton's correlation in terms of particles, it is still likely that Newton, could he have come back to life, would have raised many objections. Waves do not exist in the abstract but are waves in and of a medium, like air or water. Proponents of wave theory in fact assumed that there had to be some light-wave medium, traditionally called the *ether,* that completely filled otherwise empty space. To Newton's mind impulse or wave theories could not give reasonable evidence of the existence of this ether or explain why it had certain basic wave forms but not others and why refraction and other optical operations could not change one basic wave form into another (as can happen with air and water waves).[6]

In the twentieth century scientists have come to accept what is called wave-particle duality: under certain circumstances light appears particulate (photonic), under others it appears wave-like, and there is no longer thought to be a need for postulating an underlying ether medium. Photons are interpreted more or less as tiny particles that serve as ultimate units of electromagnetic energy, so this appears to rehabilitate, at least to a degree, Newton's conception of light as an aggregate of discrete corpuscles. Moreover, as we shall see, Newton also attributed certain wave-like characteristics to rays in order to explain the colors of thin films like soap bubbles. But we must be careful not to bend too far backward in an effort to vindicate Newton's conception of light, for twentieth-century theories go far beyond anything he imagined and are based on the quite different principles of quantum physics.

In a chapter making the transition from the letter to the *Opticks* it may seem odd to spill so much ink over the question of how later generations interpreted Newton and his theory and how we might interpret him today, yet in fact it makes it easier to see that Newton and his opponents were not simply blind to the merits of their adversaries' theories but were defending different ideals of science. Newton believed in no uncertain terms that the goal of natural philosophy was the truth about nature. If about some things we could have no more than probability through hypotheses, that was all right, but only if we also recognized that there were important things that we could really *know,* in the sense that once known was forever known. Most of Newton's opponents were inclined to think that researchers had to be willing to tolerate a high degree of uncertainty, especially since human fallibility always made it likely that some evidence or some possibility had been overlooked, and they were not ready to accept Newton's sharp differentiation between theory (proved truth) and hypothesis.

In every age since then it has been possible to identify both "certaintist" and "probabilist" attitudes toward science, to give these two tendencies abstract names. In the twentieth century scientists are more likely to be "probabilists" than "certaintists," at least when they are asked outright whether we really know what we claim to know, whereas in the period of natural science dominated by classical physics, roughly from 1700 to 1900, there would have been a much higher proportion of certaintists. Yet both tendencies probably exist not just in every age but also in every individual, and likewise there is probably a propensity in each and every researcher to prefer real truth to good hypothesis. Correspondingly, our conception of what science is will be affected both by what we consider to be our very best knowledge and by how we achieved it, and the standards erected in this light will create a certain tension with those areas where the facts of the matter are more uncertain and the approaches less well defined. That is, there will always continue to be opportunity for disputes over the "facts," the methods we use to attain them, and the interpretations we apply.

Keeping these aspects in mind, we can say that Newton generally set a high standard for experiments and experimental description ("the facts"). In many cases he gives a most painstaking account of the exact circumstances of experiments. Yet at other times his accounts are rough and ready, even oversimplified. His instructions are sometimes so vague that a reader trying to reproduce the phenomenon feels left at sea. A person performing the experiments will also notice that the phenomena are much harder to describe accurately than would appear at first glance and that Newton's descriptions often leave something to be desired (for example, a spectrum produced according to his specifications is likely to have an elongated pear or teardrop shape rather than to appear as a rectangle bounded by two semicircles). And if one tries to make all the measurements that Newton specifies, one will (1) conclude that Newton must have had astounding dexterity, speed, and experimental skill and (2) suspect that he "improved" some of them with the help of his theory.

As for Newton's method and interpretation, by looking back over the structure of the letter, in particular the series of hypothesis refutations capped by the crucial experiment that is used to demonstrate the truth of differential refrangibility, one can see that the overall argument is more complex than it at first seems. The experiments used to refute other hypotheses are barely described, and the hypotheses he treats are really his own selective interpretations of them. And can the single experiment he calls crucial really establish as much as it claims? He tries to keep issues of refrangibility distinct from color, but does he fully succeed?

These and questions like them are properly scientific ones, quite as pertinent and as much a part of science today as they were three centuries ago. Questions of fact, method, and interpretation are constantly being discussed by scientists, but most of us do not recognize this because we rarely get beyond the science that appears in introductory textbooks (where the best- and longest-established parts of a science are presented in a dogmatic way) and in the popular press.

Newton was irritated and discouraged by the prolonged disputes over his theory of white light and colors. He reached the point where he threatened to keep future work to himself. But in the course of the debate he did reveal some additional results, especially in a long paper from December 1675 presenting a "Hypothesis" about light based on a study of what we call interference effects (in particular, colored rings that appear in thin transparent films). We shall see some of this work in the second book of the *Opticks*.

For almost three decades from the mid-1670s Newton was publicly silent about light and color.[7] In the late 1670s and the early 1680s he worked instead at chemistry, alchemy, theology, and mathematics, and finally he developed a comprehensive theory of force and motion, culminating in the *Principia mathematica philosophiae naturalis* (published in 1686–1687). Eventually he left Cambridge to take public office at the Royal Mint in 1696. In 1703 he was elected president of the Royal Society of London, a post he held until his death. But Newton had not forgotten his optics. By 1690 he was at work again on a treatise of optics, first in Latin, then in English, which he hoped would remove the doubts raised by critics and incorporate phenomena that had not been part of the original presentation and theory. By 1704 he was ready to issue the first edition of this treatise, a book that more than any other established the theoretical, experimental, and technical foundations of modern optics and color science: *Opticks: or, a Treatise of the Reflexions, Refractions, Inflexions and Colours of Light*.

Part 3

The *Opticks*

Chapter 5

The *Opticks,* Book I, Part I

As of the early 1670s Newton had worked out a theory of the differential refrangibility of light and a theory of color corresponding to refrangibility; he had also begun investigating and speculating about other phenomena of light. He inclined toward thinking of rays as corpuscular, but he did not dogmatically insist on this. He did insist that his theory of differential refrangibility was better than a mere hypothesis, that it had revealed certain fundamental and unshakable truths about light and color. The objections raised against it did nothing to abate this conviction; they only confirmed his belief that people who held contrary views were hard to convince and that even those who were most expert in optics had difficulty recognizing the merits and scope of his theory.

By the mid-1670s Newton had formed a plan to collect and publish the correspondence touched off by his new theory of white light and colors, a plan that may have been frustrated by the accidental destruction of his papers. He returned to the subject around 1690, at which time the composition of the *Opticks* proper can be said to begin. The work of this period of composition seems to have involved relatively little new experimentation; Newton was chiefly writing and rewriting on the basis of his past experiences with mirrors, lenses, thin films, and the like.[1] The first edition of the *Opticks* was finally published in 1704, with some mathematical treatises appended to it; the Latin translation, with additions made to the questions, or *queries,* Newton had raised in conclusion, was published in 1706, followed by a second English edition with further queries in 1717 (the third English edition [1722], the last to appear during Newton's lifetime, and the posthumous fourth English edition [1730] were virtually unchanged from the second).

The *Opticks* is divided into three books. Book I, in two parts, covers differential refrangibility (Part I) and the associated theory of white light and colors (Part II). Book II, in four parts, takes up the phenomena produced by thin and thick films (or plates, as Newton calls them) of transparent material. Book III, in a single part, briefly considers the phenomenon of diffraction (which Newton calls inflection) and then turns to a long series of wide-ranging queries about the nature of light, the interaction of light with other matter, and what these things reveal about the fundamental constitution of physical reality.

Given such scope, one might expect a grand introduction. The *Opticks* begins, however, with a simple, direct statement of purpose and a rather dry recitation of basic definitions and axioms. Here is a sample:

THE FIRST BOOK OF OPTICKS.

PART I.

My Design in this Book is not to explain the Properties of Light by Hypotheses, but to propose and prove them by Reason and Experiments: In order to which I shall premise the following Definitions and Axioms.

DEFINITIONS.

DEFIN. I. *By the Rays of Light I understand its least Parts, and those as well Successive in the same Lines, as Contemporary in several Lines.* For it is manifest that Light consists of Parts, both Successive and Contemporary; because in the same place you may stop that which comes one moment, and let pass that which comes presently after; and in the same time you may stop it in any one place, and let it pass in any other. For that part of Light which is stopp'd cannot be the same with that which is let pass. The least Light or part of Light, which may be stopp'd alone without the rest of the Light, or propagated alone, or do or suffer any thing alone, which the rest of the Light doth not or suffers not, I call a Ray of Light.

DEFIN. II. *Refrangibility of the Rays of Light, is their Disposition to be refracted or turned out of their Way in passing out of one transparent Body or Medium into another. And a greater or less Refrangibility of Rays, is their Disposition to be turned more or less out of their Way in like Incidences on the same Medium.* Mathematicians usually consider the Rays of Light to be Lines reaching from the luminous Body to the Body illuminated, and the refraction of those rays to be the bending or breaking of those lines in their passing out of one Medium into another. And thus may Rays and Refractions be considered, if Light be propagated in an instant. But by an Argument taken from the Aequations of the times of the Eclipses of *Jupiter's Satellites,* it seems that Light is propagated in time, spending in its passage from the Sun to us about seven Minutes of time:[2] And therefore I have chosen to define Rays and Refractions in such general terms as may agree to Light in both cases.

DEFIN. III. *Reflexibility of Rays, is their Disposition to be reflected or turned back into the same Medium from any other Medium upon whose Surface they fall. And Rays are more or less reflexible, which are turned back more or less easily. . . .*

. . . .

DEFIN. VII. *The Light whose Rays are all alike Refrangible, I call Simple, Homogeneal and Similar; and that whose Rays are some more Refrangible than others, I call compound, Heterogeneal and Dissimilar.* The former Light I call Homogeneal, not because I would affirm it so in all respects, but because the Rays which agree in Refrangibility, agree at least in all those their other Properties which I consider in the following Discourse.

DEFIN. VIII. *The Colours of Homogeneal Lights, I call Primary, Homogeneal and Simple; and those of Heterogeneal Lights, Heterogeneal and Compound.* For these are always compounded of the colours of Homogeneal Lights; as will appear in the following Discourse.

AXIOMS.

AX. I. *The Angles of Reflexion and Refraction, lie in one and the same Plane with the Angle of Incidence.*

AX. II. *The Angle of Reflexion is equal to the Angle of Incidence.*

. . . .

AX. VIII. *An object seen by Reflexion or Refraction, appears in that place from whence the Rays after their last Reflexion or Refraction diverge in falling on the Spectator's Eye. . . .*

As in the February 1672 letter, Newton's express concern is not to advance hypotheses but to establish the properties of light ("to propose and prove them"); he will do this by reasoning and experiment. There is what appears to be a slight difference of emphasis, however; whereas the letter, following the method of Francis Bacon, was a narrative account of Newton's experience of prismatic colors and his gradual realization of their theoretical implications, the *Opticks* begins with a series of definitions and axioms that place the work firmly in the context of previous geometrical optics and that also present some of the key theoretical terms (e.g., homogeneal and heterogeneal lights and colors) that Newton had added to that tradition in his work before 1704.

Why the difference in the style of presentation? One thing to consider is that the *Opticks* is in outward structure like the *Principia,* which had appeared seventeen years earlier. This latter work also begins with definitions, proceeds to "axioms, or laws of motion," and then develops the mathematics of motion in accordance with those definitions and axioms in a series of lemmas, theorems, and problems: in Book I, assuming a force that decreases in effectiveness in direct proportion to the square of distance, the mathematics of motion subject only to such forces is developed; in Book II, there is added the complication of resistance to such motion by a surrounding fluid (e.g., motion in water or air instead of in a vacuum); finally, in Book III, this mathematical structure is applied to the "system of the world," that is, it is shown to describe accurately the motions of the planets, moon, and sun and the motions of bodies on earth. In a sense the first two books of the *Principia* constitute a hypothesis: suppose that bodies are mathematically idealized objects, that they stay at rest or move in a straight line unless a force is applied to them, that they interact with equal and opposite reactions, and so forth; moreover, suppose each body attracts every other body according to an inverse-square law; then how would such bodies move? This is to employ the hypothetico-deductive method, which idealizes a situation, works out the rational and mathematical consequences of the idealization, and then checks to see whether the ideal consequences correspond to what happens in the real world.

Appearances to the contrary notwithstanding, Newton rejects the axiomatic approach in the *Opticks*. If it truly paralleled the *Principia,* the first book or two would develop an idealized mathematical theory of light, and only thereafter would it be applied to real light. As we shall presently see, the *Opticks* does not develop a mathematical theory but from the beginning presents experiments tending to show ("proving") that light is differentially refrangible, that white light is composed of colors, that some of these colors are fundamental and unchangeable, and so forth. The intent might be described as more inductive than deductive. Yet even this would be somewhat misleading, because induction derives generalizations from evidence, whereas Newton has already embedded theoretical presuppositions in the definitions and axioms and in the very terms he uses to describe the evidence. And just as he is about to begin

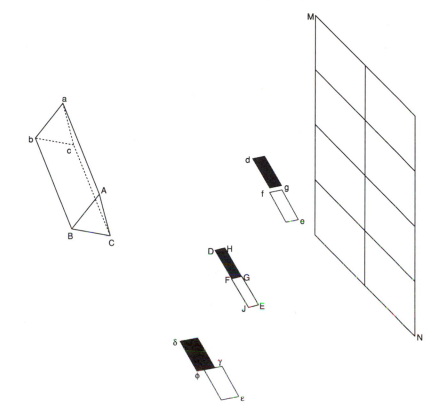

Fig. 17

experimentation he inextricably joins the notions of light, ray, and color in the reader's understanding:

PROPOSITIONS.

PROP. I. THEOR. I. *Lights which differ in Colour, differ also in Degrees of Refrangibility.*

The PROOF by Experiments.

Exper. 1 [Fig. 17]. I took a black oblong stiff Paper terminated by Parallel Sides, and with a Perpendicular right Line drawn cross from one Side to the other, distinguished it into two equal Parts. One of these parts I painted with a red colour and the other with a blue. The Paper was very black, and the Colours intense and thickly laid on, that the Phaenomenon might be more conspicuous. This Paper I view'd through a Prism of solid Glass, whose two Sides through which the Light passed to the Eye were plane and well polished, and contained an Angle of about sixty degrees; which Angle I call the refracting Angle of the Prism. And whilst I view'd it, I held it and the Prism before a Window in such manner that the Sides of the Paper were parallel to the Prism, and both those Sides and the Prism were parallel to the Horizon, and the cross Line was also parallel to it: and that the Light which fell from the Window upon the Paper made an Angle with the Paper, equal to that Angle which was made with the same Paper by the Light reflected from it to the Eye. Beyond the Prism was the Wall of the Chamber under the Window covered over with black Cloth, and the Cloth was involved in Darkness that no Light might be reflected from thence, which in passing by the Edges of the Paper to the Eye, might mingle itself with the Light of the Paper, and obscure the Phaenomenon thereof. These things being thus ordered, I found that if the refracting Angle of the Prism be turned upwards, so that the Paper may seem to be lifted upwards by the Refraction, its blue half will be lifted higher by the Refraction than its red half. But if the refracting Angle of the Prism be turned downward, so that the Paper may seem to be carried lower by the Refraction, its blue half will be carried something lower thereby than its red half. Wherefore in both Cases the Light which comes from the blue half of the Paper through the Prism to the Eye, does in like Circumstances suffer a greater Refraction than the Light which comes from the red half, and by consequence is more refrangible.

The February 1672 letter had begun with Newton getting a prism and amusing himself with the spectacle of colors. The *Opticks,* to the contrary, gets right down to business with directed experiments. The circumstances are carefully specified and, where possible, measured with some precision. Another contrast is that the *Opticks* commences with an experiment that has you looking through a prism, rather than projecting images on a screen. For Newton this was a return to the origins of his study of light: the early notebooks indicate that Newton's first prism experiments were of this kind.[3] In a sense, the first experiment is a refutation of the experimental objection raised by Anthony Lucas that was mentioned in the previous chapter. Lucas had observed the refraction of light coming from colored strips immersed in water to see which color would appear first as he peered over the edge of the basin. Here Newton eliminates possible disturbing effects of background light by using black paper as the ground. He shows that the blue patch is refracted more than the red, yet because the colors are not spectral—they reflect not just their own kind but also a lesser quantity of all others as well—

there are fringes that might induce a "careless" observer to conclude that they are really refracted the same amount.

The second experiment no longer uses a prism but instead asks the reader/experimenter to form the focused images of the blue and red patches using a lens. Because the points of focus are about 1½ inches apart, it is again clear that there is a significant difference in the refrangibility of blue-making and red-making rays.

Newton from the outset uses terms like "ray" and "refrangibility," about which he had been more cautious in the letter. In the definitions he describes rays as the smallest parts of light, whether light is propagated instantaneously or in time, whether it is corpuscular or impulse-like. Refrangibility is the property of a ray of being refracted to a certain degree when it encounters a different transparent medium. The use of these terms from the outset at least predisposes the reader to Newton's theory, and, as was pointed out in the previous chapters, it in effect decides against conceiving light as a pulse or wave phenomenon.

The entire structure of the *Opticks,* apart from the abstract definitions and axioms, is calculated to induce the reader to cooperative activity (at least mental activity, if not actual experimental activity): it is learning by doing, a how-to approach. The structure is designed to bring the reader to a destination (working with, understanding, and accepting the theory of differential refraction) rather than to lead him or her through a process of rapidly eliminating leading theories and eventually arriving at the truth through a series of crucial experiments. It is only with the second part of Book I, when Newton turns to an elaboration of the theory of colors, that he expressly brings up competing theories for refutation (though many of the experiments of the first part are already intended to disconfirm competing explanations).

Although it is not always necessary for the reader to reproduce the experiments exactly, there are certain considerations to keep in mind when introducing variations, as Newton explains in a paragraph that discusses method:

. . . *Scholium.* The same Things succeed, notwithstanding that some of the Circumstances be varied; as in the first Experiment when the Prism and Paper are any ways inclined to the Horizon, and in both when coloured Lines are drawn upon very black Paper. But in the Description of these Experiments, I have set down such Circumstances, by which either the Phaenomenon might be render'd more conspicuous, or a Novice might more easily try them, or by which I did try them only. The same Thing, I have often done in the following Experiments: Concerning all which, this one Admonition may suffice. Now from these Experiments it follows not, that all the Light of the blue is more refrangible than all the Light of the red: For both Lights are mixed of Rays differently refrangible, so that in the red there are some Rays not less refrangible than those of the blue, and in the blue there are some Rays not more refrangible than those of the red: But these Rays, in proportion to the whole Light, are but few, and serve to diminish the Event of the Experiment, but are not able to destroy it. For, if the red and blue Colours were more dilute and weak, the distance of the Images would be less than an Inch and a half; and if they were more intense and full, that distance would be greater, as will appear hereafter. These Experiments may suffice for the Colours of Natural Bodies. For in the Colours made by the Refraction of Prisms, this Proposition will appear by the Experiments which are now to follow in the next Proposition.

This scholium reflects Newton's years of experience in research and debate but also reaffirms what he had said to opponents years before. In confirming a theoretical statement some experiments work better than others, although (if the theory is true) all relevant experiments will provide evidence for the theory. In general Newton tries to provide meticulous detail about how one may reproduce especially revealing experiments.

It is quite reasonable to expect that some experiments will provide clearer evidence than others, yet this principle cannot eliminate all disagreements over the interpretation of experiments, and it could be invoked in a way that amounts to assuming the theory one is trying to verify. The scholium itself is a case of this. It explains that the light reflected by the blue paper is not all more refrangible than the light reflected by the red; rather, both lights are mixtures of differently refrangible rays. This of course is to explain the experiment by the very principle that the experiment is supposed to prove (the logical fallacy called *petitio principii,* begging the question). Yet it would perhaps be misunderstanding what Newton intended if we dismissed this as a simple fallacy. By the 1700s Newton doubtless had become so accustomed to his ray conception that it was scarcely possible for him to entertain seriously any alternative. And, as his opening statement put it, his goal was still essentially that of the letter of February 1672: not to create hypotheses by which the properties of light might be explained but instead to offer them to view (propose them) and to prove them.

"Prove" is an ambiguous term. In usage common around 1700 it could mean to demonstrate to be true beyond doubt, but it also indicated the more modest "test out," "confirm," "verify." The one meaning can shade off into the other, and in fact it seems likely that Newton intended something of both. He is not hypothetically advancing a fictional explanation but rather using experiments to exhibit fundamental characteristics (properties) of light itself and to verify that these characteristics exist and truly belong to light. The *Opticks* is *demonstration science,* a kind of scientific presentation that aims to make its theoretical points by showing especially revealing phenomena.

In the third experiment Newton proceeds to prism experiments in the fashion of the letter by admitting a beam of sunlight into a darkened room and refracting it onto a wall or screen, but the *Opticks* does a better job of informing the reader about the significance of many details. He first makes sure that the prism is in the position of minimum deviation (by rotating the prism to the point where the spectrum stops and then reverses its motion) and then describes the image's shape and dimensions. Noting that if the prism is rotated into different positions the image can be made longer or shorter, he explains the advantages of the position of minimum deviation. He describes the dimensions that result with prisms having slightly different refracting angles (64, 62½, 63½ degrees) and also with a water prism.

> . . . Neither did the matter of the Prisms make any [sensible changes]: for in a Vessel made of polished Plates of Glass cemented together in the shape of a Prism and filled with Water, there is the like Success of the Experiment according to the quantity of the Refraction. It is farther to be observed, that the Rays went on in right Lines from the Prism to the Image, and therefore at their very going out of the Prism had all that Inclination to one another from which the length of the Image proceeded, that is, the Inclination of more than two degrees and an half. And yet according to the Laws of Opticks vulgarly received, they could not possibly be so much inclined to one another.

He notes that "this Image or Spectrum PT was coloured, being red at its least refracted end T, and violet at its most refracted end P, and yellow green and blue in the intermediate Spaces. Which agrees with the first Proposition, that Lights which differ in Colour do also differ in Refrangibility." Next, in experiment 4, he looks through the prism at the aperture in the window shut, with the same kind of result, that the violet

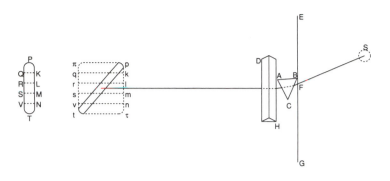

Fig. 18

part is most refracted, the red least, and so forth, contrary to the expectation raised if there were "one certain Proportion of the Sines of Incidence and Refraction."

. . . So then, by these two Experiments it appears, that in equal Incidences there is a considerable inequality of Refractions. But whence this inequality arises, whether it be that some of the incident Rays are refracted more, and others less, constantly, or by chance, or that one and the same Ray is by Refraction disturbed, shatter'd, dilated, and as it were split and spread into many diverging Rays, as *Grimaldo* [Grimaldi; see Chapter 9] supposes, does not yet appear by these Experiments, but will appear by those that follow.

Exper. 5. Considering therefore, that if in the third Experiment [which presents the refraction of light in camera obscura by a single prism] the Image of the Sun should be drawn out into an oblong Form, either by a Dilatation [stretching out] of every Ray, or by any other casual inequality of the Refractions, the same oblong Image would by a second Refraction made sideways be drawn out as much in breadth by the like Dilatation of the Rays, or other casual inequality of the Refractions sideways, I tried what would be the Effects of such a second Refraction. For this end I ordered all things as in the third Experiment, and then placed a second Prism immediately after the first in a cross Position to it, that it might again refract the beam of the Sun's Light which came to it through the first Prism.

What he found was that the spectrum now appeared along a diagonal to the crossed prisms, with the colors exactly in the expected order (see Fig. 18). He even used three and four prisms to try to cause casual or irregular "dilatations" of every ray, but the spectrum always appeared as

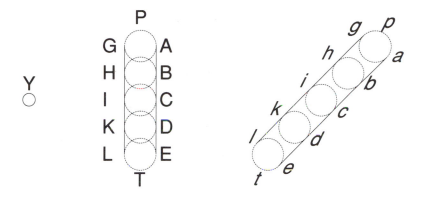

Fig. 19

expected. In part what is at issue here is the proper way to conceive of what is happening to the rays of light when refraction takes place. If the mechanism of change takes the beam and stretches it out, then, since this stretching takes place in the *direction of deviation,* successive refractions at right angles to one another should elongate the beam in different directions, directions perpendicular to one another, with the result a square or rectangle. But multiple cross refractions do not produce an elongation of a type different from that of a single refraction.[4*] Newton's alternative strategy is to use the theory of differential refrangibility according to ray type in order to picture what happens to the component rays in the original beam. The key to the explanation he is about to give is that for any single kind of ray a single sine law of refraction holds. Thus with the prism at or near minimum deviation, sun rays of a given refrangibility (and color) will produce a circular image after refraction. The entire spectrum can be treated as myriad overlapping circles, each one corresponding to rays of slightly different refrangibility:

... But that the meaning of this Experiment may more clearly appear, it is to be considered that the Rays which are equally refrangible do fall upon a Circle answering to the Sun's Disque. For this was proved in the third Experiment. By a Circle I understand not here a perfect geometrical Circle, but any orbicular Figure whose length is equal to its breadth, and which, as to Sense, may seem circular. Let therefore AG [in Fig. 19] represent the Circle which all the most refrangible Rays propagated from the whole Disque of the Sun, would illuminate and paint upon the opposite Wall if they were alone; EL the Circle which all the least refrangible Rays would in like manner illuminate and paint if they were alone; BH, CJ, DK, the Circles which so many intermediate sorts of Rays would successively paint upon the Wall, if they were singly propagated from the Sun in successive order, the rest being always intercepted; and conceive that there are other intermediate Circles without Number, which innumerable other intermediate sorts of Rays would successively paint upon the Wall if the Sun should successively emit every sort apart. And seeing the Sun emits all these sorts at once, they must all together illuminate and paint innumerable equal Circles, of all which, being according to their degrees of Refrangibility placed in order in a continual Series, that oblong Spectrum PT is composed which I described in the third Experiment.

The mechanism of refraction Newton advances appears to accord perfectly with the phenomenon. Unlike the modificationists, Newton does not assign priority to elongation but rather explains elongation as a result of something more fundamental, the composition of ordinary light out of many different kinds of light, each of which is refracted without elongation, but the sum of whose kinds produces an elongated spectrum. This is the ultimate phenomenon in Newton's way of conceiving refraction: it is the foundation of his way of theoretical seeing.

It is likewise the foundation of the experimental technique of the *Opticks*. What refraction does is to separate the kinds of light from one another; the more perfectly separated into its components white light is, the more perfectly the components exhibit their true character and behave according to unvarying law. The separated component obeys the same sine law of refraction no matter how many times it is refracted; it produces one and only one color that cannot be altered by any further experimental manipulation. It is a simple, basic entity. There is thus a premium put on isolating these basic entities: they are the fundamental constituents of light, they obey the laws of optics and color without exception, they are the ultimate confirmation of Newton's theory, and they are those things into which light must be analyzed if it is to be understood scientifically (or philosophically, as Newton and his contemporaries would have said). Thus it is not surprising that in the fourth proposition Newton turns to the problem of how to improve the separation from one another of the different components of light.[5]

... *PROP.* IV. PROB. I. *To separate from one another the heterogeneous Rays of compound Light.*

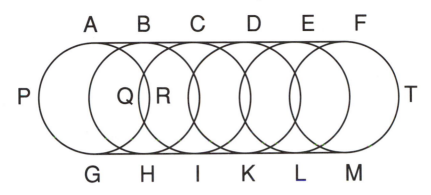

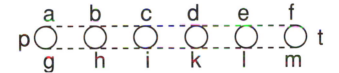

Fig. 20

The heterogeneous Rays are in some measure separated from one another by the Refraction of the Prism in the third Experiment, and in the fifth Experiment, by taking away the Penumbra[6] from the rectilinear sides of the coloured Image [by using a lens], that Separation in those very rectilinear sides or straight edges of the Image becomes perfect. But in all places between those rectilinear edges, those innumerable Circles there described, which are severally illuminated by homogeneal Rays, by interfering with one another, and being every where commix'd, do render the Light sufficiently compound. But if these Circles, whilst their Centers keep their Distances and Positions, could be made less in Diameter, their interfering one with another, and by Consequence the Mixture of the heterogeneous Rays would be proportionally diminish'd [see Fig. 20]. . . .

Now he that shall thus consider it, will easily understand that the Mixture is diminished in the same Proportion with the Diameters of the Circles. If the Diameters of the Circles whilst their Centers remain the same, be made three times less than before, the Mixture will be also three times less; if ten times less, the Mixture will be ten times less, and so of other Proportions. . . . And hence it easily follows, that the Mixture of the Rays in the refracted Spectrum *pt* is to the Mixture of the Rays in the direct and immediate Light of the Sun, as the breadth of that Spectrum is to the difference between the length and breadth of the same Spectrum.[7]

So then, if we would diminish the Mixture of the Rays, we are to diminish the Diameters of the Circles. Now these would be diminished if the Sun's Diameter to which they answer could be made less than it is, or (which comes to the same Purpose) if without Doors [outside], at a great distance from the Prism towards the Sun, some opake Body were placed, with a round hole in the middle of it, to intercept all the Sun's Light, excepting so much as coming from the middle of his Body could pass through that Hole to the Prism. For so the Circles AG, BH, and the rest, would not any longer answer to the whole Disque of the Sun, but only to that Part of it which could be seen from the Prism through that Hole, that . . . is to the apparent Magnitude of that Hole view'd from the Prism. But that these Circles may answer more distinctly to that Hole, a Lens is to be placed by the Prism to cast the Image of the Hole, (that is, every one of the Circles AG, BH, &c.) distinctly upon the Paper at PT, after such a manner, as by a Lens placed at a Window, the Species of Objects abroad are cast distinctly upon a Paper within the Room, and the rectilinear Sides of the oblong Solar Image in the fifth Experiment became distinct without any Penumbra. If this be done, it will not be necessary to place that Hole very far off, no not beyond the Window. . . .

The conception of light as composite and the spectrum as the sum of superimposed circles of simple light provides the key to solving the problem of separation. Each colored circle is in effect an image of the sun, an image made by simple light or, more properly, an image produced by the small part of the sun's light that the aperture admits into the darkened room. The smaller that aperture is made, the smaller will be the circles of light, and thus the more quickly they will be separated by refraction. How large the circles will be depends on the divergence of the extreme boundaries of the beam. A single aperture cannot limit this divergence because of the effective diameter of the sun's disk, but two or more apertures can considerably reduce the divergence. Moreover, a lens can be used to eliminate the divergence (i.e., to make the rays of the beam parallel) or at least bring the rays of a single kind to a focus, where the circle they make will be reduced to a minimum.

But then why should one deal only in circular images? The privilege assigned to the circle is an accident of the fact that the sun is circular (and also that it is easier to make a small aperture in the form of a circle, for example by using a nail to punch a hole in wood or metal, than to make a square or rectangular aperture with very straight and even sides); by

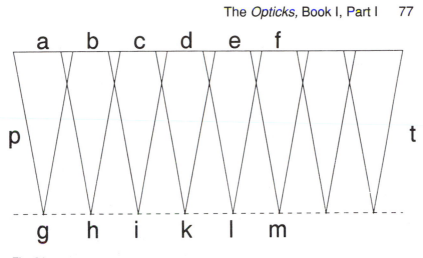

Fig. 21

using a differently shaped aperture one can get different, indeed better, results. An aperture with straight-line boundaries produces an image that lends itself to more efficient separation than does a circle. Because a circle is widest across its diameter, in any spectrum formed with a circular hole the overlapping of different kinds of light will be greatest at the center of the spectrum and progressively less toward either edge. But a narrow rectangular slit has a constant width across its entire length, so the spectrum produced from it will have a more regular and easily determinable degree of separation at every point. By using a narrow triangular slit Newton produces a spectrum that is decreasingly composite as one moves from one side of the spectrum to the other; he even claims that on the side corresponding to the narrow vertex of the triangle the separation is perfect (along line *ghiklm* in Fig. 21).

In Proposition V Newton makes explicit that "Homogeneal Light is refracted regularly without any Dilatation splitting or shattering of the Rays, and the confused Vision of Objects seen through refracting Bodies by heterogeneal Light arises from the different Refrangibility of several sorts of Rays"; the first part of this assertion, he says, was already quite well proved by the fifth experiment, but now he meets more demanding standards of separation to show that one can get "perfectly circular" images of homogeneal light by using a small second aperture to select well-separated light from a first spectrum and refracting it with a second prism. Proposition VI goes on to make explicit the claim that there is a (slightly different) sine proportion governing each of the different kinds of homogeneal light; thus the sine law is preserved not for light as a whole but for each kind apart from other kinds.

It was noted earlier that Newton first announced his theory of light and colors in 1672 to justify the design of the catoptric, or reflecting, telescope he had earlier submitted to the Royal Society of London. Newton concludes the first part of the first book of the *Opticks* with a discussion of the insuperable difficulties presented by refracting, or dioptric, telescopes and how the reflecting design overcomes these. The basic thesis is that his predecessors had unwittingly pursued a futile line of investigation by seeking to grind the surfaces of lenses to shapes other than spherical. Descartes had already shown around 1630 that a lens ground in an elliptical or hyperbolic shape (which, along with circular and parabolic curves, are called *conic sections,* because they can be obtained by cutting a cone with a plane) could eliminate the error in refraction induced by spherical curvature; the really difficult problem thus became practical, how to devise a machine or technique that could reliably grind such a shape.

In effect Newton now trumps Descartes and his confrères in optics twice: (1) by showing that if there were no difference in the refrangibility of the differently colored rays the problem could have been solved by a compound lens—that is, a lens compounded out of "sublenses" made of different substances with differing refractive indices, all ground to appropriate spherical shapes—and (2) by showing that the difference in refrangibility of the different colors induces an uncorrectable error much larger than that caused by the *spherical* shape of lenses; chromatic aberration far exceeds spherical aberration. Therefore Newton's predecessors might have saved themselves considerable theoretical and practical trouble if they had known of his theory of light. Although there were formidable problems still to be resolved in the design of effective reflecting telescopes, they were not accompanied by any theoretical impossibilities.[8*]

In Proposition VII, Newton ascertains by calculation and experiment that the chromatic aberration is more than five thousand times greater than the spherical aberration. Such a proportion tends to strain belief, Newton admits, since the obvious question is how his predecessors could have failed to notice such an astonishingly large aberration. How is it, Newton asks, that things nevertheless appear quite sharp in existing refracting telescopes? The answer Newton presents is an ingenious theory about where in the image the rays are concentrated and which colors of light the eye is most responsive to. Although this theory is not acceptable by contemporary standards, it does point to the need in a *comprehensive* theory of light and color for a theory of perceptibility (this is not Newton's term, though it is constructed in analogy to "refrangibility"), that is, a psychophysical theory of color that correlates physical events with psychological perceptions and responses.

. . . But you will say, if the Errors caused by the different Refrangibility be so very great, how comes it to pass, that Objects appear through Telescopes so distinct as they do? I answer, 'tis because the erring Rays are not scattered uniformly over all that Circular Space, but collected infinitely more densely in the Center than in any other Part of the Circle, and in the Way from the Center to the Circumference, grow continually rarer and rarer, so as at the Circumference to become infinitely rare; and by reason of their Rarity are not strong enough to be visible, unless in the Center and very near it. . . .

The "infinitely more dense" concentration of rays in the center is perhaps a bit exaggerated. Compared to the periphery of the image, where the number of rays falls off at some point to zero, the number of rays at the center is "infinitely more," but it is still not entirely plausible that the image should be as sharp as it appears, since at every point of the image there would be a "blurring" produced by the infinite variety of kinds of light coming from different points of the object being viewed. So Newton continues to develop his explanation in terms of what he calls the different luminosity of colors.

But it's farther to be noted, that the most luminous of the Prismatick Colours are the yellow and orange. These affect the Senses more strongly than all the rest together, and next to these in strength are the red and green. The blue compared with these is a faint and dark Colour, and the indigo and violet are much darker and fainter, so that these compared with the stronger Colours are little to be regarded. The Images of Objects are therefore to be placed, not in the Focus of the mean refrangible Rays, which are in the Confine of green and blue, but in the Focus of those Rays which are in the middle of the orange and yellow; there where the Colour is most luminous and fulgent [bright], that is in the brightest yellow, that yellow which inclines more to orange than to green. And by the Refraction of these Rays (whose Sines of Incidence and Refraction in Glass are as 17 and 11) the Refraction of Glass and Crystal for Optical Uses is to be measured. Let us therefore place the Image of the Object in the Focus

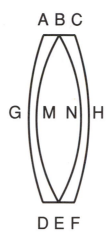

Fig. 22

of these Rays, and all the yellow and orange will fall within a Circle, whose Diameter is about the 250th Part of the Diameter of the Aperture of the Glass [of the lens].

. . . .

Now were it not for this different Refrangibility of Rays, Telescopes might be brought to a greater perfection than we have yet describ'd, by composing the Object-glass of two Glasses with Water between them. Let ADFC [in Fig. 22] represent the Object-glass composed of two Glasses ABED and BEFC, alike convex on the outsides AGD and CHF, and alike concave on the insides BME, BNE, with Water in the concavity BMEN. Let the Sine of Incidence out of Glass into Air be as I to R, and out of Water into Air, as K to R, and by consequence out of Glass into Water, as I to K: and let the Diameter of the Sphere to which the convex sides AGD and CHF are ground be D, and the Diameter of the Sphere to which the concave sides BME and BNE, are ground be to D, as the Cube Root of KK − KI to the Cube Root of RR − RI:[9] and the Refractions on the concave sides of the Glasses, will very much correct the Errors of the Refractions on the convex sides, so far as they arise from the sphericalness of the Figure. And by this means might Telescopes be brought to sufficient perfection, were it not for the different Refrangibility of several sorts of Rays. But by reason of this different Refrangibility, I do not yet see any other means of improving Telescopes by Refractions alone, than that of increasing their lengths, for which end the late Contrivance of *Hugenius* [Huygens] seems well accommodated. For very long Tubes are cumbersome, and scarce to be readily managed, and by reason of their length are very apt to bend, and shake by bending, so as to cause a continual trembling in the Objects, whereby it becomes difficult to see them distinctly: whereas by his Contrivance the Glasses are readily manageable, and the Object-glass being fix'd upon a strong upright Pole becomes more steady.

Seeing therefore the Improvement of Telescopes of given lengths by Refractions is desperate; I contrived heretofore a Perspective by Reflexion, using instead of an Object-glass a concave Metal. . . .

Newton goes on to describe the design of reflectors and, in the final proposition of Book I, Part I, how telescopes might be made shorter.

Conclusion

What Newton has achieved in the first part of the *Opticks* is the extension of the law of sines by the principle of differential refrangibility of the rays that ordinarily compose white light. He takes the optics inherited from the first part of the seventeenth century, as transmitted to him by Isaac Barrow, and brings it to a kind of completion. Earlier optical researchers had not realized that they had failed to isolate and identify the fundamental characteristic of light that Newton named refrangibility, which is another way of saying that they had failed to isolate and identify the ultimate constituents, the ultimate rays, of light. Although Newton still does not claim to have identified the nature of those constituents—for instance, not wanting to mix the certain with the conjectural, he resists asserting the corpuscular nature of light—he does claim that he has identified an ultimate and inalienable quality or characteristic of light, one that is unchangeable by experimental manipulation of the rays and that therefore deserves to be considered as fundamental to the ultimate rays, whatever they might turn out to be.

Although Newton has not been particularly reluctant to talk about colors in this first part, neither has he attempted to give a thoroughgoing discussion of them. Instead, he has focused on establishing differential refrangibility. Although this property grounds a new way of looking at and manipulating light, it is consistent with the basic geometrical approach taken by his predecessors. At the same time it permits him to establish that the premier practical problem those predecessors had sought to solve, to perfect the image-forming capabilities of lenses by finding an ideal curvature, was in fact unsolvable, and to propose the alternative of reflection as far superior. But, as the rest of the *Opticks* shows, Newton has a much larger agenda for the science of light than the first part indicates. Taking on that agenda requires first settling the difficult problem that had puzzled many of the best minds before him: the problem of color.

Chapter 6

The *Opticks,* Book I, Part II

After the extended discussion of refrangibility of Book I, Part I, Newton turns immediately to the other major question raised by his experiments, the nature of color.

THE FIRST BOOK OF OPTICKS.

PART II.

PROP. I. THEOR. I. *The Phaenomena of Colours in refracted or reflected Light are not caused by new Modifications of the Light variously impress'd, according to the various Terminations of the Light and Shadow.*

It is interesting, in light of the letter of February 1672, that Newton begins his treatment of color in the *Opticks* in an adversarial or polemical mode. His first impulse is not to establish something positive about color but to remove an opposing, in fact the hitherto *predominant*, view about the nature of color: that it was caused by some modification of light.

The Proof by Experiments.
Exper. 1. For if the Sun shine into a very dark Chamber through an oblong hole F [in Fig. 23], whose breadth is the sixth or eighth part of an Inch, or something less; and his beam FH do afterwards pass first through a very large Prism ABC, distant about 20 Feet from the hole, and parallel to it, and then (with its white part) through an oblong hole H, whose breadth is about the fortieth or sixtieth part of an Inch, and which is made in a black opake Body GI, and placed at the distance of two or three Feet from the Prism, in a parallel Situation both to the Prism and to the former hole, and if this white Light thus transmitted through the hole H, fall afterwards upon a white Paper *pt,* placed after that hole H, at the distance of three

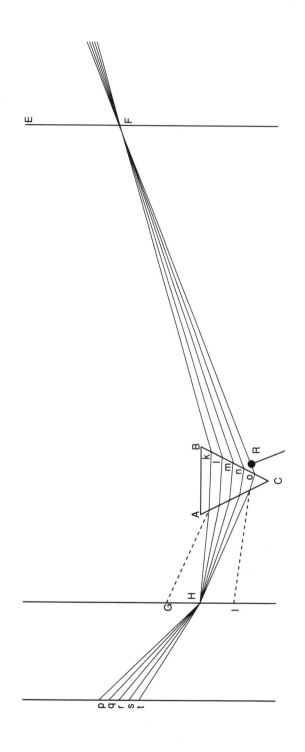

Fig. 23

or four Feet from it, and there paint the usual Colours of the Prism, suppose red at *t,* yellow at *s,* green at *r,* blue at *q,* and violet at *p;* you may with an Iron Wire, or any such like slender opake Body, whose breadth is about the tenth part of an Inch, by intercepting the Rays at *k, l, m, n* or *o,* take away any one of the Colours at *t, s, r, q* or *p,* whilst the other Colours remain upon the Paper as before; or with an Obstacle something bigger you may take away any two, or three, or four Colours together, the rest remaining: So that any one of the Colours as well as violet may become outmost in the Confine of the Shadow towards *p,* and any one of them as well as red may become outmost in the Confine of the Shadow towards *t,* and any one of them may also border upon the Shadow made within the Colours by the Obstacle R intercepting some intermediate part of the Light; and, lastly, any one of them by being left alone, may border upon the Shadow on either hand. All the Colours have [behave] themselves indifferently to any Confines of Shadow, and therefore the differences of these Colours from one another, do not arise from the different Confines of Shadow, whereby Light is variously modified, as has hitherto been the Opinion of Philosophers. . . .

The natural philosophers who advanced modification theories of color cited various hypothetical mechanisms that were supposed to generate colors at the edges of a refracted beam of light, that is, at the places where light was bounded by surrounding darkness. In the present experiment, Newton attempts to show that the colors of the spectrum appear in an order that is indifferent to the location of bounding obstacles. He now uses rectangular apertures rather than circular to avoid the complications that arise when the degree of mixing of the light varies between the edges and the center of the spectrum. A small rectangular aperture picks out a segment of the exterior sunlight and admits it into the darkened room. At the distance of 20 feet a prism refracts the beam toward a black sheet of paper just a few feet further on, in which there is cut a rectangular slit, through which a small portion of the refracted light passes on to the wall or screen. By putting a thin wire or similar obstacle in the beam just before it reaches the prism, he is able to "remove" from the spectrum on the wall any color or colors he chooses. In particular, violet can be removed so that blue is the extreme color; green can be removed to produce a divided spectrum, with violet and blue extremities in the one half and yellow and red in the other; and so forth. If the colors of the spectrum were formed by an edge disturbance at the boundaries between light and dark, all spectra and divided spectra presumably would have violet at one end and red at the other, contrary to the result of this experiment.[1*]

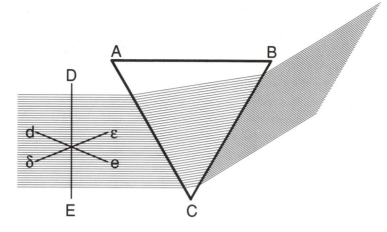

Fig. 24

In the next two experiments Newton gives a different proof to show that boundaries are not responsible for the colors. Experiment 3 is easier to perform, because unlike 2 (which is otherwise similar) it does not require a lens. Admit a broad beam of light into a darkened room, refract it immediately, and then, just beyond the prism, take an unlined white notecard and use it as a screen. When it is held perpendicular to the direction of the light's path you will see a white image fringed with color. Now progressively rotate the card around an axis parallel to the axis of the prism, first in one direction, then in the other (positions *de* and δε in Fig. 24). You will see the light on the card gradually become completely yellow and red in the one case, blue and violet in the other. In this ingenious experiment, the color in the refracted light is made to change without any additional bounding of the light (unlike the first experiment, where the wire and the two narrow apertures bound it). The basis for Newton's explanation is the theory of differential refrangibility: the least-refrangible and most-refrangible rays depart from the second face of the prism at slightly different angles, so that a white screen held at a very oblique angle will be illuminated by color-producing rays that strike it ever so glancingly while the rays that produce other colors pass by completely unintercepted because they are parallel to the plane of the screen or card.

Experiment 4 introduces a different refutation of the claim that boundaries cause color:

Exper. 4. The Colours of Bubbles with which Children play are various, and change their Situation variously, without any respect to any Confine or Shadow. If such a Bubble be cover'd with a concave Glass, to keep it from being agitated by any Wind or Motion of the Air, the Colours will slowly and regularly change their Situation, even whilst the Eye and the Bubble, and all Bodies which emit any Light, or cast any Shadow, remain unmoved. And therefore their Colours arise from some regular Cause which depends not on any Confine of Shadow. What this Cause is will be shewed in the next Book.

The only thing that is changing here is the bubble (the thickness of the bubble wall); all other circumstances remain the same. There is no boundary to be identified here, yet the bubble produces changing colors. (Actually, even in this case there is a boundary: a beam of light is "self-bounding," since its limits are in essence defined by the boundary between where there is brightness and where there is none. Nevertheless, it appears once again that the modificationists have no explanation.)[2*] And, as Newton remarks, this experiment also anticipates the subject matter of the second book of the *Opticks*, the colors produced by thin transparent films, the theory of which will be used to explain the colors produced by ordinary bodies.

Newton now turns to an elaboration of the positive theory of colors that is implied by the theory of differently refrangible rays.

. . . *PROP.* II. THEOR. II. *All homogeneal Light has its proper Colour answering to its Degree of Refrangibility, and that Colour cannot be changed by Reflexions and Refractions.*

In the Experiments of the fourth Proposition of the first Part of this first Book, when I had separated the heterogeneous Rays from one another, the Spectrum *pt* formed by the separated Rays, did in the Progress from its End *p*, on which the most refrangible Rays fell, unto its other End *t,* on which the least refrangible Rays fell, appear tingod with this Series of Colours, violet, indigo, blue, green, yellow, orange, red, together with all their intermediate Degrees in a continual Succession perpetually varying. So that there appeared as many Degrees of Colours, as there were sorts of Rays differing in Refrangibility.

The proposition says, first, that with homogeneal light color corresponds to refrangibility and, second, that this color of homogeneal light is unchangeable. The first clause is "proved" by referring the reader back to Book I, Part I. The second clause will be tested by the experiments that follow.

The theory of refrangibility of Book I, Part I argued that there is an indefinitely, perhaps infinitely large number of different kinds of rays. This is one point in the theory where the claim goes well beyond the evidence. For instance, it is conceivable that the number, though very large, is small enough so that if one reduced the cross section of the physical ray to a far greater degree than Newton did, one would find a discontinuous spectrum, with very tiny but detectable gaps. The reader probably knows that the spectrum of sunlight contains gaps corresponding to wavelengths that have been absorbed by atoms in the atmospheres of the sun and the earth. Newton never observed such lines.[3] But it is at least interesting to wonder what he might have made of them and whether they would have caused him to alter his theory of light. At the very least it might have induced him to wonder whether there really is a different kind of ray corresponding to every conceivable mathematical difference in degree of refrangibility. This is simply to point out that the visible evidence has its limits, that it cannot justify in any ultimately decisive way a conclusion that goes beyond what has been experimentally verified (or what is experimentally verifiable).

There is a related problem when we shift from questions of refrangibility to those of color. Newton contends that there are as many degrees of color as there are degrees of refrangibility. But regardless of the kind of spectrum one looks at (whether produced according to Newton's instructions or with a modern spectroscope), the number of distinct hues will be finite, even if one separates small neighboring segments of a spectrum and compares them one against one. With a complete spectrum one sees broad fields of more or less uniform hue separated by narrow regions of rapid color transition. It is even plausible to claim that one sees a segment of fairly uniform violet next to a narrower segment of fairly uniform indigo next to a broader one of blue, and so forth, finally to a broad patch of fairly uniform red. In a sense, one sees just the five to seven colors that are ordinarily identified and not all the intermediate gradations corresponding to indefinitely numerous kinds of rays. Those who have heard that there are millions of hues in the spectrum will find it surprising how few can actually be discriminated there.

Consider a solar spectrum produced by modern prisms and diffraction gratings. Typically, the difference in wavelengths of two narrow lines of light must be between 40 (in the violet) and 90 (in the red) ten-billionths of a meter (or *angstroms*) in order for the hues to be discriminable under the most favorable viewing conditions. Since, even generously conceived, the visible wavelengths of radiation range over less than 4,000 angstroms, if one takes just one hundred spectral lines at more or less equal distances from one another it will be extremely difficult to distinguish the hue of any one of them from its immediate neighbors to the right and left.[4] A major source of the discrepancy between the theoretical claim of millions of hues and the actual phenomena is that such large estimates are really not about the solar spectrum at all. They of course include the hues of the spectrum of bright sunlight but also (1) all colors produced by adding white light to spectral colors, which addition results in hues less intense than the corresponding spectral hues, (2) all mixtures of two or more spectral colors that produce hues not identical to those in the pure spectrum, and (3) all differences produced by the greater or lesser brightness of the light used to produce the spectrum (leading to greater or lesser saturation in the spectral colors). The point is that a particular solar spectrum consists of light at a determinate brightness, without additional white light being mixed in, and relatively unmixed (depending on the degree of separation) with wavelengths other than those immediately proximate.

How important is this discrepancy between the number of colors predicted theoretically and the number seen? The answer depends on one's purposes and one's standards of judgment. To some critics of Newton (like Hooke), Newton had unnecessarily multiplied the kinds of entities in order to explain the phenomena. Others might be inclined instead to dismiss the discrepancy as more apparent than real, based on oversubtle philosophical arguments or a misplaced belief that every objective difference in the ray will correspond to a difference in perception. It is likely that in this case a judicious verdict should seek a middle ground while noting that there are very important issues at stake on this point.

First, as the February 1672 letter revealed, Newton considered the theory of refrangibility and the theory of colors to be very closely related. The theory of refrangibility depends heavily for its proof on the observer's ability to discriminate the quality of color, just as much as on the more quantitative determination of the degree to which a ray of light has been turned out of its original path by refraction. Insofar as the basic theory of color is questioned, so is also an important "evidentiary prop" of the theory of refrangibility. Newton pronounced his goal in the *Opticks* as that of proposing and proving certain properties of light by reason and experiments; how certain and convincing the proof can be hinges on these things. Clearly, Newton did not intend us to read the *Opticks* as simply a presentation of a very good hypothesis; he meant it as a true account of the properties, among others, of refrangibility and color.

Second, to use an apt metaphor, Newton is looking at colors through the prism of his theory of differential refrangibility. His ingenious notion of conceiving the spectrum as overlapping circles plus the discovery that once-refracted, spectrally colored light gives different results from white light[5]—in particular, subsequent refractions do not spread out the light in the way that the original white light is spread and do not materially alter the color if one is careful about making the light as homogeneal as possible—lead naturally to the idea that refrangibility and color are not just closely correlated but perhaps even results of a common cause, a single underlying reality. It is therefore quite conceivable that color is just as objective a quality as is refrangibility.

Such convictions can quite easily lead to the contention that in some sense the spectral colors are more real, or truer, colors than colors produced by "mixed" light (including white). The real nature of color thus would go hand in hand with refrangibility, and to every characteristic in the one there would have to be a corresponding characteristic in the other. In particular, refrangibility is a property of light that is easily turned into a mathematical quantity; the strong correlation between refrangibility and color implies that color is mathematizable as well. This might induce anyone to seize upon those characteristics of the color phenomena that are most easily mathematized and to neglect others that are not. This is, in effect, Newton's way of seeing and treating color theoretically.

The fault here, if it may be called one, is in trying to establish a too perfect correspondence between perception and the physics of radiation. In the past 150 years it has been recognized that establishing the correlations between radiation and perceived color is a complex matter.[6] As is often the case, however, the earlier generations of investigators made, expressly or implicitly, simplifying assumptions that facilitated their work but also required later revision. Newton deserves credit for having opened up the discipline of color psychophysics, but his work is properly subject to revision and criticism, especially where it is clear that he overstepped the bounds of what his evidence supported.

The next experiments are devoted to confirming that homogeneal colors are unchangeable by subsequent experimental manipulation; this makes it possible to take them as a foundation for the theory of how mixed colors are produced.

Exper. 5. Now, that these Colours could not be changed by Refraction, I knew by refracting with a Prism sometimes one very little Part of this Light, sometimes another very little Part. . . . If any Part of the red Light was refracted, it remained totally of the same red Colour as before. No orange, no yellow, no green or blue, no other new Colour was produced by that Refraction. Neither did the Colour any ways change by repeated Refractions, but continued always the same red entirely as at first. The like Constancy and Immutability I found also in the blue, green, and other Colours. . . . All Bodies illuminated with compound Light appear through Prisms confused, (as was said above) and tinged with various new Colours, but those illuminated with homogeneal Light appeared through Prisms neither less distinct, nor otherwise colour'd, than when viewed with the naked Eyes. Their Colours were not in the least changed by the Refraction of the interposed Prism. I speak here of a sensible Change of Colour: For the Light which I here call homogeneal, being not absolutely homogeneal, there ought to arise some little Change of Colour from its Heterogeneity. But, if that Heterogeneity was so little as it might be made by the said Experiments of the fourth Proposition, that Change was not sensible, and therefore in Experiments, where Sense is Judge, ought to be accounted none at all.

Newton is saying that if the color is truly homogeneal the refraction cannot change the color; if there is some very slight change, or if it is hard to tell whether there is one, then the light used might not have been perfectly homogeneal. A more rigorous control of the experiment might eliminate the apparent error. And the more perfectly homogeneal the light is, the sharper the objects illuminated by it will appear when viewed through a prism.

Yet the latter part of the passage seems a bit convoluted: one might react by saying that either the color is changed by refraction or it is not, and the talk of "sensible change" just muddies the waters. In response to Newton's assertion that any small sensible change can be eliminated by doing things more rigorously, a critic could argue that telling us what *would* happen if we did something we do not actually do is not to show that that is how things *really* turn out. The passage is perhaps again evidence of theory guiding seeing. On the other hand, this also shows the way in which theory justifiably focuses our attention and gives us ways of determining what is an acceptable margin of error. If employing more sophisticated techniques of isolating light enables us to reduce the error till it is hardly noticeable, we feel justified in extrapolating to what might happen if we could actually reach theoretical perfection in practice.

Exper. 6. And as these Colours were not changeable by Refractions, so neither were they by Reflexions. For all white, grey, red, yellow, green, blue, violet Bodies . . . in red homogeneal Light appeared totally red, in blue Light totally blue, in green Light totally green, and so of other Colours. In the homogeneal Light of any Colour they all appeared totally of that same Colour, with this only Difference, that some of them reflected that Light more strongly, others more faintly. I never yet found any Body, which by reflecting homogeneal Light could sensibly change its Colour.

From all which it is manifest, that if the Sun's Light consisted of but one sort of Rays, there would be but one Colour in the whole World, nor would it be possible to produce any new Colour by Reflexions and Refractions, and by consequence that the variety of Colours depends upon the Composition of Light.

To put the existence of rays to one side: if there were not something about sunlight that could be differentiated, then the light reflected by any body whatsoever would be the same except for quantity, and it would be impossible to understand how there could be the diversity of colors in such undifferentiated stuff. Something like this would be true regardless of the theory of color to which you hold. For example, modificationists believe that white light is the most basic kind of light and that something in its nature allows it to be changed (modified) so that it can display colors.

There follows immediately a passage that makes a distinction of vital importance: Are the rays of light themselves colored?

DEFINITION. The homogeneal Light and Rays which appear red, or rather make Objects appear so, I call Rubrifick or Red-making; those which make Objects appear yellow, green, blue, and violet, I call Yellow-making, Green-making, Blue-making, Violet-making, and so of the rest. And if at any time I speak of Light and Rays as coloured or endued with Colours, I would be understood to speak not philosophically and properly, but grossly, and accordingly to such Conceptions as vulgar People in seeing all these Experiments would be apt to frame. For the Rays to speak properly are not coloured. In them there is nothing else than a certain Power and Disposition to stir up a Sensation of this or that Colour. For as Sound in a Bell or musical String, or other sounding Body, is nothing but a trembling Motion, and in the Air nothing but that Motion propagated from the Object, and in the Sensorium [the place of sensation] 'tis a Sense of that Motion under the Form of Sound; so Colours in the Object are nothing but a Disposition to reflect this or that sort of Rays more copiously than the rest; in the Rays they are nothing but their Dispositions to propagate this or that Motion into the Sensorium, and in the Sensorium they are Sensations of those Motions under the Forms of Colours.

Thus light and light rays cannot properly be called colored; they have not color but a tendency to produce color when they illuminate visible objects. There is a *disposition* in the rays to *produce* colors. The "form" of colors properly appears in conscious sensation, which Newton terms the *sensorium*. To put it another way, color is not ultimately a property of either the object we call colored or the light that reveals it as such but a property of consciousness; the object, in its turn, has the property of reflecting this or that kind of light in such-and-such a proportion, while the light has the property of producing colors.

In effect, this is an affirmation by Newton of the distinction between primary and secondary qualities. In principle the distinction had already been enunciated in late sixteenth-century Italy (something like it existed even in Greek antiquity, especially among atomist and Stoic philosophers), and it can be found in influential modern form in the writings of Galileo, Descartes, and Hobbes. Robert Boyle, a natural philosopher a generation senior to Newton, seems to have given the distinction its name, and it was further popularized by the philosopher John Locke (1632-1704).[7] Primary qualities are those that truly belong to the thing in question and that produce ideas that closely resemble them in consciousness. For example, the physical dimensions of a cube really belong to that cube and are perceived as they truly are. The list of primary qualities usually includes at least size, shape, and motion. Secondary qualities are those that, although often due to some arrangement of matter's primary qualities, arouse ideas that do not actually resemble the cause. Thus an object can produce sensations such as tickling, pain, color, taste, and aroma, but these perceived qualities do not truly belong to the object. Color may be caused by the effect of atoms on photons; yet what we see is not atoms, photons, atom-photon interactions, or the microscopic structure of matter, but the appearance of red, green, or blue.

It could be claimed that the primary-secondary distinction was essential to the rise of modern science because it forms the basis for the distinction between objective and subjective knowledge. Be that as it may, it is in effect a critical response to the traditional Aristotelian-Scholastic conception of qualities and perception. A Scholastic formula had it that the sensible in act was the same as the sense in act: that is, the quality of the thing that was capable of being sensed was put into action by something else (in the case of color by the presence of light in the medium between the thing and the eye), and this action was conveyed to the sense organ, which thereupon shared the actuality of the thing; indeed, the actuality or activity of the sense organ was the same as the actuality or activity of the thing.

A puzzle that this conception left unresolved was the precise manner of conveyance of actuality from the object to the sense organ. If one pictures it in terms of some kind of transmission in a material reality (like an ether through which an impulse or pressure is carried, or particles that fly at incredible speed) it becomes difficult to understand how someone can argue that an aspect of the thing is transmitted, integrally and without any significant change, to the organ. It seems more plausible to assume that the thing interacts in some way with the intermediating matter, which is not usually of the same nature as the original thing, and the result of this interaction is transmitted, by pulse or particles, to yet a third thing, the sense organ, where it again interacts with other things (nerves, animal spirits, the brain) in a way that ultimately gets interpreted, consciously or unconsciously, by the mind.[8]

Newton rejects the Aristotelian conception, yet it is not certain that he is completely in sympathy with the primary-secondary distinction. For one thing, he does not assert, either dogmatically or philosophically, that the fundamental characteristics of the world are extension, motion, location, and the like (though one might want to claim that his mathematical approach implies something like this). For another, the tenor of Newton's theory of color, both here and in the February 1672 letter, holds out the possibility that color is either itself mathematical or so intimately connected with something mathematical that color is intrinsically mathematizable. Insofar as the primary-secondary distinction implies that the primary qualities reveal something proper to the thing itself whereas the secondary only reveal the character of the interaction of a hidden property of the thing with the perceiver, Newton seems to hesitate. Perceived colors, at least in certain circumstances, stand in a very close connection with fixed characteristics of rays of light and, as his theory of thin-plate colors will try to show, with the fundamental arrangement of matter in bodies, too.

Having established the principles of the theory of color in general, Newton now turns to a more precise and mathematical determination of color. By drawing lines to separate the different fields of color in the spectrum, he comes to the conclusion that the colors are arranged in the same way as the tones in the musical scale.

PROP. III. PROB. I. *To define the Refrangibility of the several sorts of homogeneal Light answering to the several Colours.*

For determining this Problem I made the following Experiment.

Exper. 7. When I had caused the Rectilinear Sides AF, GM, [in Fig. 25] of the Spectrum of Colours made by the Prism to be distinctly defined, as in the fifth Experiment of the first Part of this Book is described, there were found in it all the homogeneal Colours in the same Order and Situation one among another as in the Spectrum of simple Light, described in the fourth Proposition of that Part. For the Circles of which the Spectrum of compound Light PT is composed, and which in the middle Parts of the Spectrum interfere, and are intermix'd with one another, are not intermix'd in their outmost Parts where they touch those Rectilinear Sides AF and GM. And therefore, in those Rectilinear Sides when distinctly defined, there is no new Colour generated by Refraction. I observed also, that if any where between the two outmost Circles TMF and PGA a Right Line, as γδ, was cross to the Spectrum, so as both Ends to fall perpendicularly upon its Rectilinear Sides, there appeared one and the same Colour, and degree of Colour from one End of this Line to the other. I delineated therefore in [on] a Paper the Perimeter of the Spectrum FAP GMT, and in trying the third Experiment of the first Part of this Book, I held the Paper so that the Spectrum might fall upon this delineated Figure, and agree with it exactly, whilst an Assistant, whose Eyes for distinguishing Colours were more critical than mine, did by Right Lines αβ, γδ, εζ, &c. drawn cross the Spectrum, note the Confines of the Colours. . . . And this Operation being divers times repeated both in the same, and in several Papers, I found that the Observations agreed well enough with one another, and that the Rectilinear Sides MG and ΓA were by the said cross Lines divided after the manner of a Musical Chord. Let GM be produced [extended] to X, that MX may be equal to GM, and conceive GX, λX, ιX, ηX, εX, γX, αX, MX, to be in proportion to one another, as the Numbers, 1, 8/9, 5/6, 3/4, 2/3, 3/5, 9/16, 1/2, and so to represent the Chords of the Key [the tonic], and of a Tone [a whole tone above the tonic], a third Minor, a fourth, a fifth, a sixth Major, a seventh and an eighth [octave] above that Key: And the intervals Mα, αγ, γε, εη, ηι, ιλ, and λG, will be the Spaces which the several Colours (red, orange, yellow, green, blue, indigo, violet) take up.

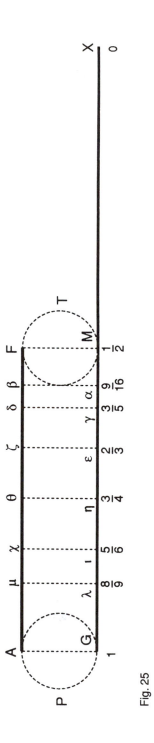

Fig. 25

Newton's choices for the proportions characteristic of the musical intervals in the seven-tone, or diatonic, scale would produce small dissonances (to our ears) because they are based on *just intonation* rather than on our practice of tempering the scale. Just intonation was introduced in the sixteenth century to eliminate marked dissonances in the scale built on perfect fifths, the Pythagorean scale (named after the ancient Greek philosopher-mathematician Pythagoras). In this scale the ratio of the length of a string representing the tonic note to the length of another string (of the same thickness and at the same tension) representing the fifth above it, for example the relationship between C and G, is exactly 3 to 2. If you take a series of twelve fifths above a given note, say a C, you should reach another C seven octaves higher. Octaves are perfect consonances with the ratio 2:1, so the lower C should be represented by a string $2^7 = 128$ times the length of the higher C's string. But if you calculate the ratio by the twelve superimposed perfect fifths, you get $(3/2)^{12} = 531,441/4,096 = 129.746+$, a difference of approximately 1.5%. Just intonation overcomes this discrepancy by making a compromise: two slightly different whole tones are used at different points in the scale, one expressed by the ratio 9/8 (type 1 whole tone), the other by 10/9 (type 2 whole tone). Newton's compromise scale is as follows: tonic note, up a type 1 whole tone (since the string for the higher tone is shorter, the proportion is actually the inverse of 9/8, that is, 8/9), up a half tone (for which the proportion is 16/15, so that the previous length, $1 \times 8/9$, is in turn multiplied by the inverse, 15/16), up a type 2 whole tone (multiply the previous result by the inverse of 10/9), up a type 1 whole tone, up a type 2 whole tone, up a half tone, up a type 1 whole tone. This results in the sequence of relative lengths 1, 8/9, 5/6, 3/4 (a perfect fourth), 2/3 (a perfect fifth), 3/5, 9/16, and the octave 1/2. This scale is basically that of the white notes beginning on D.[9]

Now these Intervals or Spaces subtending the Differences of the Refractions of the Rays going to the Limits of those Colours, that is, to the Points M, α, γ, ϵ, η, ι, λ, G, may without any sensible Error be accounted proportional to the Differences of the Sines of Refraction of those Rays having one common Sine of Incidence, and therefore since the common Sine of Incidence of the most and least refrangible Rays out of Glass into Air was (by a Method described above) found in proportion to their Sines of Refraction, as 50 to 77 and 78, divide the Difference between the Sines of Refraction 77 and 78, as the Line GM is divided by those Intervals, and you will have 77, 77 ⅛, 77 ⅕, 77 ⅓, 77 ½, 77 ⅔, 77 ⅞, 78, the

Sines of Refraction of those Rays out of Glass into Air, their common Sine of Incidence being 50. . . .

The division of the spectrum interval from 77 to 78 accords with the musical proportions mentioned in the previous paragraph. Double all the numbers in that sequence (doubling does not change the proportional relations between the members of the sequence) to get: 2×1, $2 \times \frac{8}{9}$, $2 \times \frac{5}{6}$, and so forth, that is, 2, $1\frac{7}{9}$, $1\frac{2}{3}$, $1\frac{1}{2}$, $1\frac{1}{3}$, $1\frac{1}{5}$, $1\frac{1}{8}$, 1. If we read the sequence backwards, we see that the points of fractional division are the same as those between 77 and 78.

Newton's efforts to establish a connection between the diatonic musical scale and the colors in the spectrum could be accounted an oddity reminiscent of number mysticism, a quest for strange and even far-fetched analogies like those to be found a century earlier, for instance in the work of the great astronomer and optical scientist Johannes Kepler. Besides, the attempt was misguided, for, contrary to what Newton thought, the portions of the spectrum occupied by the different colors are not the same for all refracting materials, so that no single, authoritative measurements like these can be given.

Yet there *are* strange and surprising analogies to be found in nature, so that no one should be criticized for seeking them. As we know today, both light and sound can be understood as wave phenomena, something that is by no means clear from surface appearances. Historically, there has been a tradition going back at least to the Pythagoreans of ancient Greece that perception is based on ratios—a theory suggested and partially confirmed precisely by the theory of musical harmonies. Newton's interest in this tradition and in harmonic theory goes back at least to his student days at Cambridge (the 1660s). He first published the idea of a spectrum scale in his 1675 "Hypothesis," although it is already present in the earlier optical lectures.[10] As we shall see in the account of Book II of the *Opticks,* he tried to extend this notion of color harmonics to the phenomena of thin-plate colors. And in the Scholium to Book III, Proposition VIII of the *Principia* he went so far as to argue that Pythagoras himself understood the inverse square law that relates the attraction between the planets and sun to their distance from the sun by analogy with the proportions in the musical scale. This, in turn, gives a glimpse into Newton's belief that the fundamental truths about nature had been known to the wise men of antiquity and the Biblical prophets, who (he thought) had chosen to conceal their knowledge from the "vulgar" by using symbols and allegory.

It would lead us rather far afield to pursue this theme further (although it quite clearly is present in Book III of the *Opticks,* in Query 31), but even just mentioning it helps make clear that for Newton the spectrum scale was not simply an ingenious idea that came (so to speak) out of the blue. As one scholar has noted, in the optical lectures of the early 1670s Newton first explored the spectral proportions in terms of just the five colors red, yellow, green, blue, and violet; he then deliberately added orange and indigo in order to get a reasonably good match with the system of just musical intonation.[11] The spectrum scale was therefore not so much a happy, accidental discovery as an adjustment of the data to make them demonstrate an analogy between sound and vision.

Even taking color on its own terms, Newton must have recognized that a theory of color based on the mathematics of the refrangibility of rays was not fully satisfying as long as regularities evident in the phenomena were not reflected in the mathematics. There is no reason, initially, to expect that the infinite mathematical possibilities for degrees of refrangibility (say, between extreme red rays with index 1.54 and extreme violet with index 1.56) will be grouped into seven subclasses corresponding to the distinct areas of the ROYGBIV colors, yet that is the way the spectrum appears. One thing Newton may have suspected is that a certain similarity between the extremes of red and violet, which are bridged in nature by the nonspectral purples, suggests that the interval from red to violet corresponds more or less to an octave (the octave, of course, represents a similarity of sound at double and half the sound-wave frequency).

After finishing his account of this musical analogy, Newton draws a more general conclusion: that he has established the foundations of a new, mathematical way of treating color phenomena comprehensively, including those that are produced by mixtures of different kinds of rays. That is, he has a theory not just of spectral or homogeneal colors but of all colors produced by light.

And these Theorems being admitted into *Opticks,* there would be scope enough of handling that Science voluminously after a new manner; not only by teaching those things which tend to the perfection of Vision, but also by determining mathematically all kinds of Phaenomena of Colours which could be produced by Refractions. For to do this, there is nothing else requisite than to find out the Separations of heterogeneous Rays, and their various Mixtures and Proportions in every Mixture. By this way of arguing I invented almost all the Phaenomena described in these Books, beside some others less necessary to the Argument; and by the successes I met with in the Trials, I dare promise, that to him who shall argue truly, and then try all things with good Glasses and sufficient Circumspection, the expected Event will not be wanting. But he is first to know what Colours will arise from any others mix'd in any assigned Proportion.

That is, to know what colors are seen all one needs to know is how many of each kind of differently refrangible ray are to be found in the light that reaches the eye; the study of perceived color comes down to an analysis of rays, *ray analysis*. For such analysis to have any predictive power, however, we must first ascertain what colors result when you mix different quantities of different lights. To establish the principles of ray mixing, Newton turns to experiments once again.

PROP. IV. THEOR. III. Colours may be produced by Composition which shall be like to the Colours of homogeneal Light as to the Appearance of Colour, but not as to the Immutability of Colour and Constitution of Light. And those Colours by how much they are more compounded by so much are they less full and intense, and by too much Composition they may be diluted and weaken'd till they cease, and the Mixture becomes white or grey. There may be also Colours produced by Composition, which are not fully like any of the Colours of homogeneal Light.

For a Mixture of homogenal red and yellow compounds an Orange, like in appearance of Colour to that orange which in the series of unmixed prismatick Colours lies between them; but the Light of one orange is homogeneal as to Refrangibility, and that of the other is heterogeneal, and the Colour of the one, if viewed through a Prism, remains unchanged, that of the other is changed and resolved into its component Colours red and yellow. And after the same manner other neighbouring homogeneal Colours may compound new Colours, like the intermediate homogeneal ones, as yellow and green, the Colour between them both, and afterwards, if blue be added, there will be made a green the middle Colour of the three which enter the Composition. For the yellow and the blue on either hand, if they are equal in quantity they draw the intermediate green equally towards themselves in Composition, and so keep it as it were in Aequilibrio [equilibrium], that it verge not more to the yellow on the one hand, and to the blue on the other, but by their mix'd Actions remain still a middle Colour. To this mixed green there may be farther added some red and violet, and yet the green will not presently cease, but only grow less full and vivid, and by increasing the red and violet, it will grow more and more dilute, until by the prevalence of the added

Colours it be overcome and turned into whiteness, or some other Colour. So if to the Colour of any homogenal Light, the Sun's white Light composed of all sorts of Rays be added, that Colour will not vanish or change its Species, but be diluted, and by adding more and more white it will be diluted more and more perpetually. Lastly, If red and violet be mingled, there will be generated according to their various Proportions various Purples, such as are not like in appearance to the Colour of any homogeneal Light, and of these Purples mix'd with yellow and blue may be made other new Colours.

First he notes that two spectral colors A and C produce a color B′ like the one (B) that is intermediate between them, with the difference, of course, that if light B′ is passed through a prism it will be decomposed, whereas light B will be unchanged, since it is already homogeneal. Mixing three or more colors produces a more complicated result, which Newton approaches by describing how each color affects the previous sum of colors as it is added to them. The general notion here is that of a new color "drawing" the existing color "toward" itself. The metaphor suggests thinking of the spectrum as a continuum in which each color will try to drag other colors toward itself, with the proviso that as more and more colors are added, the hues become more dilute or faded, more white or rather more dirty and grayish. Finally, adding red to violet produces colors (purples) the kind of which are not seen in the original spectrum (nonspectral colors).

This result suggests something that we already know or suspect: that white itself is the result of mixing all, or a very large number of, the different kinds of rays. In the next proposition Newton turns to establishing this in detail.

PROP, V. THEOR. IV. *Whiteness and all grey Colours between white and black, may be compounded of Colours, and the whiteness of the Sun's Light is compounded of all the primary Colours mix'd in a due Proportion.*

Newton offers a variety of experiments in which he "mixes" colored lights, for example by overlapping the spectral colors from multiple prisms. He also uses a large comb-like device to intercept a part of the colored spectrum (this is a systematically developed variant of the experiment at the outset of Part II that employs a wire to intercept a part of the spectrum); the rest of the spectral light passes between the teeth and is brought to a focus with a lens. Thus he first separates the white light into its spectral components, then removes some of those components, and finally remixes what remains in order to observe the result. Finally, he tries an experiment mixing pigments rather than light:

. . . *Exper.* 15. Lastly, In attempting to compound a white, by mixing the coloured Powders which Painters use, I consider'd that all colour'd Powders do suppress and stop in them a very considerable Part of the Light by which they are illuminated. For they become colour'd by reflecting the Light of their own Colours more copiously, and that of all other Colours more sparingly, and yet they do not reflect the Light of their own Colours so copiously as white Bodies do. If red Lead, for instance, and a white Paper, be placed in the red Light of the colour'd Spectrum made in a dark Chamber by the Refraction of a Prism, as is described in the third Experiment of the first Part of this Book; the Paper will appear more lucid than the red Lead, and therefore reflects the red-making Rays more copiously than red Lead doth. And if they be held in the Light of any other Colour, the Light reflected by the Paper will exceed the Light reflected by the red Lead in a much greater Proportion. And the like happens in Powders of other Colours. And therefore by mixing such Powders, we are not to expect a strong and full White, such as is that of Paper, but some dusky obscure one, such as might arise from a Mixture of Light and Darkness, or from white and black, that is, a grey, or dun, or russet brown, such as are the Colours of a Man's Nail, of a Mouse, of Ashes, of ordinary Stones, of Mortar, of Dust and Dirt in High-ways, and the like. And such a dark white I have often produced by mixing colour'd Powders. . . .

Now, considering that these grey and dun Colours may be also produced by mixing Whites and Blacks, and by consequence differ from perfect Whites, not in Species of Colours, but only in degree of Luminousness, it is manifest that there is nothing more requisite to make them perfectly white than to increase their Light sufficiently; and, on the contrary, if by increasing their Light they can be brought to perfect Whiteness, it will thence also follow, that they are of the same Species of Colour with the best Whites, and differ from them only in the Quantity of Light. And this I tried as follows. I took the third of the above-mention'd grey Mixtures, (that which was compounded of Orpiment, Purple, Bise, and *Viride Aeris*)[12] and rubbed it thickly upon the Floor of my Chamber, where the Sun shone upon it through the opened Casement; and by it, in the shadow, I laid a Piece of white Paper of the same Bigness. Then going from them to the distance of 12 or 18 Feet, so that I could not discern the Unevenness of the Surface of the Powder, nor the little Shadows let fall from the gritty Particles thereof; the Powder appeared intensely white, so as to transcend even the Paper it self in Whiteness, especially if the Paper were a little shaded from the Light of the Clouds, and then the Paper compared with the Powder appeared of such a grey Colour as the Powder had done before. But by laying the Paper where the Sun shines through the Glass of the Window, or by shutting the Window that the Sun might shine through the Glass upon the Powder, and by such other fit Means of increasing or decreasing the Lights wherewith the Powder and Paper were illuminated, the Light wherewith the Powder is illuminated may be made stronger in such a due Proportion than the Light wherewith the Paper is illuminated, that they shall both appear exactly alike in Whiteness. For when I was trying this, a Friend coming to visit me, I stopp'd him at the Door, and before I told him what the Colours were, or what I was doing; I asked him, Which of the two Whites were the best, and wherein they differed? And after he had at that distance viewed them well, he answer'd, that they were both good Whites, and that he could not say which was best, nor wherein their Colours differed. Now, if you consider, that this White of the Powder in the Sun-shine was compounded of the Colours which the component Powders (Orpiment, Purple, Bise, and *Viride Aeris*) have in the same Sun-shine, you must acknowledge by this Experiment, as well as by the former, that perfect Whiteness may be compounded of Colours.

From what has been said it is also evident, that the Whiteness of the Sun's Light is compounded of all the Colours wherewith the several sorts of Rays whereof that Light consists, when by their several Refrangibilities they are separated from one another, do tinge Paper or any other white Body whereon they fall. For those Colours (by *Prop.* II. *Part* 2.) are unchangeable, and whenever all those Rays with those their Colours are mix'd again, they reproduce the same white Light as before.

The experiments mixing light and those that mix powders and pigments Newton puts on a par. It was not until the middle of the nineteenth century that the distinction between additive and subtractive color mixing was made. Adding differently colored *lights* to one another is additive mixing, which produces increased brightness with each added component. Mixing *pigments,* on the other hand, reduces brightness, since each kind of pigment selectively absorbs light at certain wavelengths. It is not possible in general to give a single set of color-mixing rules that apply equally well to both kinds of mixing. For example, blue and yellow pigments combined produce a middle or dark green, but blue and yellow light produce white or a very whitish green. Yet the point Newton is after is that by combining a sufficient variety of hues in appropriate quantities, we can nullify the chromatic effect, with the result a white or gray. White, grays, and black thus exist on a continuum;

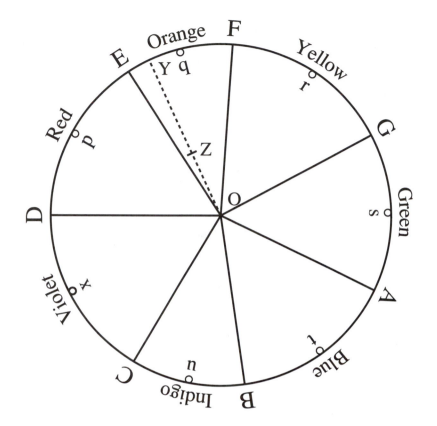

Fig. 26

all result from a balanced mixture of color pigments or color-producing rays, and the difference between the purest white and a very dark gray is solely in the quantity of light involved.[13*]

These discussions of mixing issue in the following proposition, which along with the doctrine of differential refrangibility according to color is regarded as the foundation of subsequent color theory: Newton's color-mixing circle, which serves in effect as a device to calculate what colors will result when rays of different kinds are mixed.

PROP. VI. PROB. II. In a mixture of Primary Colours, the Quantity and Quality of each being given, to know the Colour of the Compound.

With the Center O [in Fig. 26] and Radius OD describe a Circle ADF, and distinguish its Circumference into seven Parts DE, EF, FG, GA, AB, BC, CD, proportional to the seven Musical Tones or Intervals of the eight Sounds, *Sol, la, fa, sol, la, mi, fa, sol,* contained in an eight [octave], that is, proportional to the Number $\frac{1}{9}$, $\frac{1}{16}$, $\frac{1}{10}$, $\frac{1}{9}$, $\frac{1}{10}$,[14] $\frac{1}{16}$, $\frac{1}{9}$. Let the first Part DE represent a red Colour, the second EF orange, the third FG yellow, the fourth GA[15] green, the fifth AB blue, the sixth BC indigo, and the seventh CD violet. And conceive that these are all the Colours of uncompounded Light gradually passing into one another, as they do when made by Prisms; the Circumference DEFGABCD, representing the whole Series of Colours from one end of the Sun's colour'd Image to the other, so that from D to E be all degrees of red, at E the mean Colour between red and orange, from E to F all degrees of orange, at F the mean between orange and yellow, from F to G all degrees of yellow, and so on. Let *p* be the Center of Gravity[16] of the Arch [arc of the circle] DE, and *q, r, s, t, u, x,* the Centers of Gravity of the Arches EF, FG, GA, AB, BC, and CD respectively, and about those Centers of Gravity let Circles proportional to the Number of Rays of each Colour in the given Mixture be describ'd; that is, the Circle *p* proportional to the Number of the red-making Rays in the Mixture, the Circle *q* proportional to the Number of the orange-making Rays in the Mixture, and so of the rest. Find the common Center of Gravity of all those Circles *p, q, r, s, t, u, x.* Let that Center be Z; and from the Center of the Circle ADF, through Z to the Circumference, drawing the Right Line OY, the Place of the Point Y in the Circumference shall shew the Colour arising from the Composition of all the Colours in the given Mixture, and the Line OZ shall be proportional to the Fulness or Intenseness of the Colour, that is, to its distance from Whiteness. As if Y fall in the middle between F and G, the compounded Colour shall be the best yellow; if Y verge from the middle towards F or G, the compound Colour shall accordingly be a yellow, verging towards orange or green. If Z fall upon the Circumference, the Colour shall be intense and florid in the highest Degree; if it fall in the mid-way between the Circumference and Center, it shall be but half so intense, that is, it shall be such a Colour as would be made by diluting the intensest yellow with an equal quantity of whiteness; and if it fall upon the center O, the Colour shall have lost all its intenseness, and become a white. But it is to be noted, That if the point Z fall in or near the line OD, the main ingredients being the red and violet, the Colour compounded shall not be any of the prismatick Colours, but a purple, inclining to red or violet, accordingly as the point Z lieth on the side of the line DO towards E or towards C, and in general the compounded violet is more bright and more fiery than the uncompounded. Also if only two of the primary Colours which in the circle are opposite to one another be mixed in an equal proportion, the point Z shall fall upon the center O, and yet the Colour compounded of those two shall not be perfectly white, but some faint anonymous Colour. For I could never yet by mixing only two primary Colours produce a perfect white. Whether it may be compounded of a mixture of three taken at equal distances in the circumference I do not know, but of four or five I do not much question but it may. But these are Curiosities of little or no moment to the understanding the Phaenomena of Nature. For in all whites produced by Nature, there uses to be a mixture of all sorts of Rays, and by consequence a composition of all Colours.

. . . This Rule I conceive accurate enough for practice, though not mathematically accurate . . .

The color circle is a calculating device. Conceive it this way: Take a complete spectrum and bend it into a circle, with the violet end butting up against the red end. Paste this on the edge of a circular cardboard disk so that the disk is perfectly balanced at the center when it rests on a blunt, vertical needle. Suppose now that someone gives you instructions like this: take ten parts of the middle red, five of orange, seven parts middle yellow, and so forth. Your task is to find out what color this combination will produce. You get out your set of small lead weights (all equal), and attach ten on the edge of the disk at the red position, five at the orange position, seven at yellow, and so forth. Then by moving the needle around you find the new balance point, that is, the new center of gravity. Draw a line from the center of the disk through this center of gravity and note which hue it intersects at the edge. For argument's sake pretend it is green at 480 nm. What color should you see? A green, which will be pale, medium, or intense depending on the distance of the center of gravity from the center of the disk; the closer to the edge the center of gravity is, the more intense or like the spectral color it will be, the closer to the center it is the more it will be pale. The center itself is white.

This circle of color is ancestor to many descendants in two and three dimensions, though they are ordinarily used for quite different purposes today (even in the eighteenth century many of them served to describe "color space" rather than calculate hues). One of the best known is the C.I.E. (*Commission Internationale de l' Eclairage*) chromaticity diagram.[17]

At the end of the passage just quoted it looks as though Newton misses the principle of complementary colors, that is, color pairs that when combined produce white. He seems doubtful that a true white can be produced by pairs of homogeneal opposites. Since the principle of complementarity has more to do with the physiology of the visual system than with the colors of things, Newton's point about white is valid as far as it goes: in nature what counts is that white is produced by the tendency of bodies to reflect all rays equally. He is not fundamentally concerned with the "curiosities" of vision.[18*] Moreover, if he had confirmed for himself spectral complementarity he would have been faced with a puzzle: how is it possible to get a white—which by his definition is the result of a due proportion of all, or at least many, of the different kinds of rays—from just two homogeneal colors?

Although Newton recognizes that his color circle does not produce perfectly correct results, he finds it good enough for practical purposes; the sense of sight will adjudge it either quite accurate or very nearly so.

The next proposition reveals the true scope of Newton's color theory. It explains not just reflection and refraction, but all colors in nature:

PROP. VII. THEOR. V. *All the Colours in the Universe which are made by Light, and depend not on the Power of Imagination, are either the Colours of homogeneal Lights, or compounded of these, and that either accurately or very nearly, according to the Rule of the foregoing Problem.*

Clearly this proposition is not susceptible of proof; it is rather a conclusion to be drawn by extrapolation from all that has gone before. But the basic reasoning is clear: if a color is produced by light (rather than by the imagination, blows to the eye, or the like), then its proper character is determined by the ray composition of that light and is subject to the relationships expressed in the color-mixing circle. After briefly explaining this proposition, Newton turns to a detailed account of how, according to his color theory, one can explain the spectrum as being produced by a progressive separation of the rays from the moment the light leaves the prism (Proposition VIII, Problem III: "By the discovered Properties of Light to explain the Colours made by Prisms"). He also accounts for the inversion of the spectral colors that occurs when you look through a prism at the objects around you and other phenomena associated with refraction.

The next three propositions (IX–XI) bring the first book to a close. Propositions IX and X explain the rainbow and the colors of natural bodies, while XI parallels the final experiment of the letter of 32 years before by presenting a more technically sophisticated "decomposition" and then "recomposition" of the sun's light to show that it produces the same white both before and after. In Proposition X, by illuminating colored bodies with different spectral lights, Newton shows that the bodies tend to reflect abundantly spectral colors of the same kind as their natural color (e.g., red bodies reflect spectral reds), whereas they reflect little or none of other spectral colors; he promises to explain this tendency to differential *reflection* in the second book. The crowning triumph of Book I, however, is doubtless Proposition IX, the account of the rainbow.

The explanation of the rainbow was a progressive accomplishment beginning in the high Middle Ages. The state of understanding immediately before Newton was brought to its highest pitch by Descartes. This is hardly surprising, since key to an explanation is the geometric understanding of refraction, for which Descartes had enunciated the law. Rainbows are produced by two refractions and one or more reflections in essentially spherical water droplets. The primary bow is produced by a refraction as the light enters the droplet, an internal reflection at the back surface, and a second refraction out of the droplet. Sometimes a dimmer secondary and even a tertiary bow can be observed outside the primary bow; these are produced by rays entering at angles such that they are reflected twice or three times inside the droplet.

In essence the phenomenon of the rainbow is a function of the index of refraction of water and the circular or globular shape of the drop: the shape combined with the laws of refraction and reflection determines where the rainbow can be seen, because a beam of parallel rays (e.g., the sunlight that shines on the drop) is refracted by and reflected within the drop in a manner that produces a localized concentration of rays at positions that vary slightly with the index of refraction. Since Descartes considered the sine law to have a single value for all light entering a given transparent medium, he could deduce the angles at which the refracted and reflected white-producing rays were concentrated but had to explain the rainbow's colors as a *modification* of the white light, precisely the kind of explanation that Newton believed he had subverted once and for all. By providing a theory of multiple indices of refraction, one for each different kind of light ray, Newton was able to provide a single principle that explained both the shape and the colors of the rainbow.

. . . *PROP.* IX. PROB. IV. *By the discovered Properties of Light to explain the Colours of the Rain-bow.*

This Bow never appears, but where it rains in the Sun-shine, and may be made artificially by spouting up Water which may break aloft, and scatter into Drops, and fall down like Rain. For the Sun shining upon these Drops certainly causes the

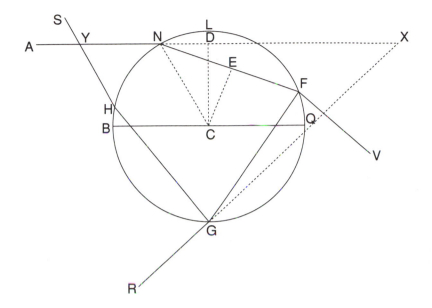

Fig. 27

Bow to appear to a Spectator standing in a due Position to the Rain and Sun. And hence it is now agreed upon, that this Bow is made by Refraction of the Sun's Light in drops of falling Rain. This was understood by some of the Antients [ancients], and of late more fully discover'd and explain'd by the famous *Antonius de Dominis* Archbishop of *Spalato,* in his Book *De Radiis Visûs & Lucis,* published by his Friend *Bartolus* at *Venice,* in the Year 1611, and written above 20 Years before. For he teaches there how the interior Bow is made in round Drops of Rain by two Refractions of the Sun's Light, and one Reflexion between them, and the exterior by two Refractions, and two sorts of Reflexions between them in each Drop of Water, and proves his Explications by Experiments made with a Phial full of Water, and with Globes of Glass filled with Water, and placed in the Sun to make the Colours of the two Bows appear in them. The same Explication *Des-Cartes* hath pursued in his Meteors,[19] and mended that of the exterior Bow. But whilst they understood not the true origin of Colours, it's necessary to pursue it here a little farther. For understanding therefore how the Bow is made, let a Drop of Rain, or any other spherical transparent Body be represented by the Sphere BNFG, [in Fig. 27] described with the Center C, and Semi-diameter [or radius] CN. And let AN be one of the Sun's Rays incident upon it at N, and thence refracted to F, where let it either go out of the Sphere by Refraction towards V, or be reflected to G; and at G let it either go out by Refraction to R, or be reflected to H; and at H let it go out by Refraction towards S, cutting the incident Ray in Y. Produce [extend] AN and RG, till they meet in X, and upon AX and NF, let fall the Perpendiculars CD and CE, and produce CD till it fall upon the Circumference at L. Parallel to the incident Ray AN draw the Diameter BQ, and let the Sine of Incidence out of Air into Water be to the Sine of Refraction as I to R. Now, if you suppose the Point of Incidence N to move from the Point B, continually till it come to L, the Arch QF will first increase and then decrease, and so will the Angle AXR which the Rays AN and GR contain; and the Arch QF and angle AXR will be biggest when ND is to CN as $\sqrt{II - RR}$ to $\sqrt{3RR}$, in which case NE will be to ND as 2R to I.[20] . . .

Although not every reader will be able to follow all the mathematical reasoning, the general principle is easy enough to grasp. The sun is to be imagined off to the left of the diagram; its rays are taken as essentially parallel to one another, and thus when they reach the surface of the droplet they are following lines parallel to BQ and AYDX. If the droplet were a cube instead of a globe there would not be any rainbow produced; since the incidence would be perpendicular to the surface the light would not be refracted, and it would simply pass through the other side. But because the droplet has a circular cross section, the angle of incidence between the points B and L varies continuously from 0° to 90°, and because of the curvature of the back edge of the droplet much of the light that enters off center will be refracted so that it strikes the back of the droplet at an angle resulting in total internal reflection. Moreover, if you trace the paths of rays entering the droplet between B and L, you find, as you proceed from B, that the point where they strike the back side moves upward away from Q, stops, and then moves back again toward Q. Assuming this point of stoppage is F, we find that more rays fall at or near F than at other points along the arc QF, and the exact position of point F varies as does the refrangibility (and thus the color) of the rays. The rays are totally reflected at this point, so they proceed toward G, where most are refracted out into the air; since the concentration of the different color-producing rays varies slightly (red-producing rays will be concentrated slightly to the right of G, violet-producing slightly to the left), we see a multicolor rainbow if we are standing along the line RGQX. The angle that this line makes with the original parallel rays depends only on the index of refraction, so the rainbow will be visible only to those who are properly situated.

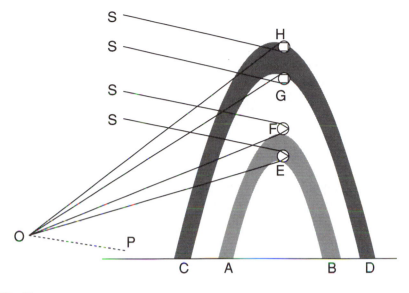

Fig. 28

Newton next calculates the angles with respect to the rays of the extremes of both the primary and secondary bows (i.e., he is determining the angles AXR and SYA for the extreme red-producing and violet-producing rays in each bow). What he shows is that the primary bow will appear between the angles of 40° 17′ (where the most refrangible or violet rays will appear) and 42° 2′ (red), and outside that bow will appear the secondary bow between the angles 50° 57′ (red) and 54° 7′ (violet). These angles can be measured either from the sun to the droplet to the eye or, alternatively, from the droplet to the eye to the point near the horizon that would be the center of the bow if the bow's curvature were extended into a full circle. The appearance of the bows is therefore a simple consequence of the mathematics of refraction and reflection in raindrops.

Suppose now that O [in Fig. 28] is the Spectator's Eye, and OP a Line drawn parallel to the Sun's Rays, and let POE, POF, POG, POH, be Angles of 40 Degr. 17 Min. 42 Degr. 2 Min. 50 Degr. 57 Min. and 54 Degr. 7 Min. respectively, and these Angles turned about their common Side OP, shall with their other Sides OE, OF; OG, OH, describe the Verges of two Rain-bows AF BE and CHDG. For if E, F, G, H, be drops placed any where in the conical Superficies described by OE,

OF, OG, OH, and be illuminated by the Sun's Rays SE, SF, SG, SH; the Angle SEO being equal to the Angle POE, or 40 Degr. 17 Min. shall be the greatest Angle in which the most refrangible Rays can after one Reflexion be refracted to the Eye, and therefore all the Drops in the Line OE shall send the most refrangible Rays most copiously to the Eye, and thereby strike the Senses with the deepest violet Colour in that Region [and, similarly, to the other angles correspond other colors]. . . .

As was noted earlier, Newton concludes by stating in Proposition X the principle that object colors depend on the propensity of bodies to reflect some kinds of rays more than others (Book II will attempt to provide further insight into the cause of this), whereas in Proposition XI he offers a more sophisticated version of the experiment that had concluded his letter of 6 February 1672, one recombining the refracted light of the sun into a new beam of white light that is then rerefracted to show that a spectrum will be produced whether the light is "naturally" or "artificially" composed out of all the different kinds of rays.

PROP. X. PROB. V. *By the discovered Properties of Light to explain the permanent Colours of Natural Bodies.*

These Colours arise from hence, that some natural Bodies reflect some sorts of Rays, others other sorts more copiously than the rest. Minium reflects the least refrangible or red-making Rays most copiously, and thence appears red. Violets reflect the most refrangible most copiously, and thence have their Colour, and so of other bodies. Every Body reflects the Rays of its own Colour more copiously than the rest, and from their excess and predominance in the reflected Light has its Colour.

. . . *PROP.* XI. PROB. VI. *By mixing colour'd Lights to compound a beam of Light of the same Colour and Nature with a beam of the Sun's direct Light, and therein to experience the Truth of the foregoing Propositions.*

Thus, in the *Opticks* the theory that was made public in 1672 has been presented in an ampler way than was possible in the February letter to the Royal Society. But Newton understood that the phenomena of refrangibility according to color did not exhaust the phenomena of light and colors. Book II therefore turns to things the letter of 1672 had left out entirely.

Chapter 7

The *Opticks,* Book II, Parts I and II

In the book *Micrographia* (1665), the man who would become Newton's lifelong nemesis, Robert Hooke, reported experiments in which by pressing plates of glass together one can observe colored rings or circles where the plates meet. Hooke attempted to measure these and to offer a rudimentary theory of how they came about, with marginal success. In December 1675, nearly four years after the appearance of the February 1672 letter, Newton published a new paper in which he first publicly addressed these thin-plate colors (which today are called thin-film colors and understood as resulting from wave interference; they are commonly seen in bubbles and in oil floating on water). Book II of the *Opticks* presents a careful, quantitative study of this phenomenon. In particular, the third part of Book II offers an explanation of them in terms of "fits of easy passage," and in turn this theory of fits is used to account for the colors of natural bodies. Since thin-plate phenomena cannot simply be reduced to refrangibility, for Newton they constitute evidence of a new property of light, and thus, in accordance with what he said in the opening statement of the *Opticks,* he proceeds to prove this property by reason and experiments.

THE SECOND BOOK OF OPTICKS.

PART I.

Observations concerning the Reflexions, Refractions, and Colours of thin transparent Bodies.

It has been observed by others, that transparent Substances, as Glass, Water, Air, &c. when made very thin by being blown into Bubbles, or otherwise formed into Plates, do exhibit various Colours according to their various thinness, altho' at a greater thickness they appear very clear and colourless. In the former Book I forbore to treat of these Colours, because they seemed of a more difficult Consideration, and were not necessary for establishing the Properties of Light there discoursed of.[1*] But because they may conduce to farther Discoveries for compleating the Theory of Light, especially as to the constitution of the parts of natural Bodies, on which their Colours or Transparency depend; I have here set down an account of them. To render this Discourse short and distinct, I have first described the principal of my Observations, and then consider'd and made use of them. The Observations are these.

Unlike in the experiments investigating refrangibility, there is no need, at least for the moment, of a darkened room, a narrow beam of sunlight, and so forth. Rather, the initial observations are made in ordinary surroundings. Newton takes two triangular prisms and presses together two sides, which he describes as slightly bulging rather than flat. Looking down at the prisms, he discovers that, depending on the angle at which he holds them, the ambient light entering the prism at the face closest to his eyes is transmitted through the spot where the prisms touch (that is, the light that enters the prism on the viewing side passes from the first prism into the second, with a resulting dark spot, because the light is not reflected back at him); but where there is air between the faces (that is, where they are not touching) some or all of the light is reflected back at the observer.

Obs. 1. Compressing two Prisms hard together that their sides (which by chance were a very little convex) might somewhere touch one another: I found the place in which they touched to become absolutely transparent, as if they had there been one continued piece of Glass. For when the Light fell so obliquely on the Air, which in other places was between them, as to be all reflected; it seemed in that place of contact to be wholly transmitted, insomuch that when look'd upon, it appeared like a black or dark spot, by reason that little or no sensible Light was reflected from thence, as from other places; and when looked through it seemed (as it were) a hole in that Air which was formed into a thin Plate, by being compress'd between the Glasses. And through this hole Objects that were beyond might be seen distinctly, which could not at all be seen through other parts of the Glasses where the Air was interjacent [between the pieces of glass]. Although the Glasses were a little convex, yet this transparent spot was of a considerable breadth, which breadth seemed principally to proceed from the yielding inwards of the parts of the Glasses, by reason of their mutual pressure. For by pressing them very hard together it would become much broader than otherwise.

In this initial observation, Newton is using familiar equipment to produce a new experimental situation. Where the sides of the prisms touch, they act as though they formed a single glass solid, letting light shining on their interface pass through (if the light is coming from above while one simultaneously observes from above, this area looks darker than the surrounding area, and through it one can see objects beyond the prisms). Where they do not touch, there is a thin film of air between the two glass surfaces. Because for many rays the angle at which the light strikes in this region is sufficiently oblique, there takes place in the upper prism the familiar phenomenon of total internal reflection. If you turn the prisms to reduce this oblique angle, eventually some of the light will stop being reflected and pass through into the "interjacent" air.

At this point Newton is not performing a carefully controlled experiment. He is simply observing, under ordinary circumstances, what happens as he turns and manipulates these prisms that are pressed together. The light reflected or let pass at the prisms' joint interface consists of rays entering the glass at various angles rather than of a sharply focused beam. Therefore, when total reflection begins occurring for some rays, it will not occur for others striking at a less oblique angle; in addition, since total internal reflection occurs sooner for violet-producing rays than for red, and since the angle at which it commences is a function of the index of refraction, we can suspect that the cause of the phenomenon has something to do with the differential refrangibility of light. But these are as yet only factors to be considered, and Newton does not even mention them. At this point he is collecting observations that are still neither sufficiently many nor sufficiently precise to justify a firm conclusion.

Continuing to turn the combined prisms so that the angle of viewing becomes more oblique, Newton proceeds to further observation:

Obs. 2. When the Plate of Air, by turning the Prisms about their common Axis, became so little inclined to the incident Rays, that some of them began to be transmitted, there arose in it many slender Arcs of Colours which at first were shaped almost like the Conchoid,[2] as you see them delineated in the [twenty-ninth] Figure. And by continuing the Motion of the Prisms, these Arcs increased and bended more and more about the said transparent spot, till they were compleated into Circles or Rings incompassing it, and afterwards continually grew more and more contracted.

These Arcs at their first appearance were of a violet and blue Colour, and between them were white Arcs of Circles, which presently by continuing the

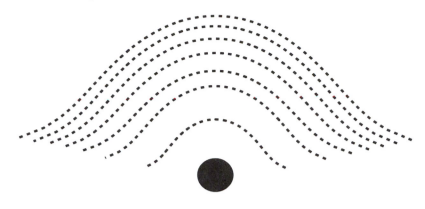

Fig. 29

Motion of the Prisms became a little tinged in their inward Limbs [edges] with red and yellow, and to their outward Limbs the blue was adjacent. So that the order of these Colours from the central dark spot, was at that time white, blue, violet; black, red, orange, yellow, white, blue, violet, &c. But the yellow and red were much fainter than the blue and violet.

The Motion of the Prisms about their Axis being continued, these Colours contracted more and more, shrinking towards the whiteness on either side of it, until they totally vanished into it. And then the Circles in those parts appear'd black and white, without any other Colours intermix'd. But by farther moving the Prisms about, the Colours again emerged out of the whiteness, the violet and blue at its inward Limb, and at its outward Limb the red and yellow. So that now their order from the central Spot was white, yellow, red; black; violet, blue, white, yellow, red, &c. contrary to what it was before.

After reporting how many successions of colored rings he was able to produce and ways of improving the circumstances of observation, Newton exchanges prisms for lenses, which, because they are ground to a precise, continuously varying curvature, allow for an exact measurement of the width of the layer of air where each color is produced.

. . . *Obs.* 4. To observe more nicely the order of the Colours which arose out of the white Circles as the Rays became less and less inclined to the Plate of Air; I took two Object-glasses,[3] the one a Plano-convex [flat on one side, bulging on the other] for a fourteen Foot Telescope, and the other a large double Convex for one of about fifty Foot; and upon this, laying the other with its plane side downwards, I pressed them slowly together, to make the Colours successively emerge in the

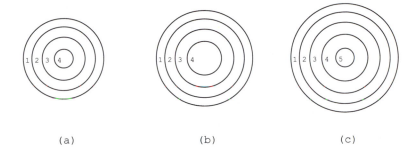

(a)   (b)   (c)

Fig. 30. The pattern of emergence of the colored rings as two lenses are pressed together with increasing force. (a) There appears within the central circle 3 a new circle (4) of different hue very small at first; (b) this new circle 4 increases in size along with the circles (or rather annuli or rings) surrounding it; (c) yet another hue emerges at the center in the form of a small circle (5), and the cycle begins again.

middle of the Circles, and then slowly lifted the upper Glass from the lower to make them successively vanish again in the same place. The Colour, which by pressing the Glasses together, emerged last in the middle of the other Colours, would upon its first appearance look like a Circle of a Colour almost uniform from the circumference to the center, and by compressing the Glasses still more, grow continually broader until a new Colour emerged in its center, and thereby it became a Ring encompassing that new Colour. And by compressing the Glasses still more, the diameter of this Ring would increase, and the breadth of its Orbit or Perimeter decrease until another new Colour emerged in the center of the last [see Fig. 30]: And so on until a third, a fourth, a fifth, and other following new Colours successively emerged there, and became Rings encompassing the innermost Colour, the last of which was the black Spot. And, on the contrary, by lifting up the upper Glass from the lower, the diameter of the Rings would decrease, and the breadth of their Orbit increase, until their Colours reached successively to the center; and then they being of a considerable breadth, I could more easily discern and distinguish their Species than before. And by this means I observ'd their Succession and Quantity to be as followeth.

The technique here is to press together lenses in order to observe the progressive generation of a series of colors, each of which emerges as a central circle or disk; as the lenses are pressed harder a new color emerges at this disk's center, turning the former disk into a ring or donut shape (the technical term for which is an *annulus*), which at first has

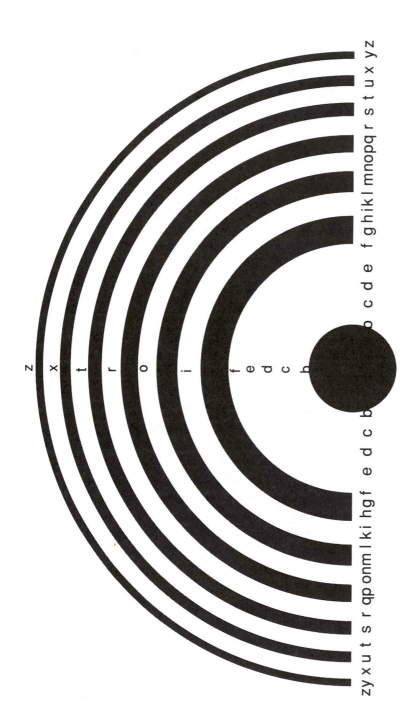

Fig. 31. A schematic illustration of Newton's rings. The letters indicate the positions of the hues described by Newton in observation 4 of Book II, Part I. The dark, hatched circle at the center is the black or dark area where the two pieces of glass touch, allowing the light to pass without reflection. Moving outward from this circle, white rings indicate the positions of the brighter hues (reds, oranges, yellows), whereas dark rings indicate the darker hues (greens, blues, violets, purples). The outermost white area of the figure, at z and beyond, is white in fact. The spacing of the rings follows the mathematical formulas Newton specifies rather than the spacing in the figure that appeared in the original edition of the *Opticks*.

wide borders that get narrower as the whole ring expands from the center. Because the lenses, unlike the prisms, initially make contact at what is virtually a geometric point, the black central hole (which transmits all light) does not appear until a whole series of rings has been generated by increasing pressure.

Having arrived at a controlled technique for generating the colored rings, Newton describes in detail what a full set of rings looks like, from the central dark spot outward.

Next to the pellucid [transparent] central Spot made by the contact of the Glasses succeeded blue, white, yellow, and red. The blue was so little in quantity, that I could not discern it in the Circles made by the Prisms, nor could I well distinguish any violet in it, but the yellow and red were pretty copious, and seemed about as much in extent as the white, and four or five times more than the blue. The next Circuit in order of Colours immediately encompassing these were violet, blue, green, yellow, and red: and these were all of them copious and vivid, excepting the green, which was very little in quantity, and seemed much more faint and dilute than the other Colours. Of the other four, the violet was the least in extent, and the blue less than the yellow or red. The third Circuit or Order was purple, blue, green, yellow, and red; in which the purple seemed more reddish than the violet in the former Circuit, and the green was much more conspicuous, being as brisk and copious as any of the other Colours, except the yellow, but the red began to be a little faded, inclining very much to purple. After this succeeded the fourth Circuit of green and red. The green was very copious and lively, inclining on the one side to blue, and on the other side to yellow. But in this fourth Circuit there was neither violet, blue, nor yellow, and the red was very imperfect and dirty. Also the succeeding Colours became more and more imperfect and dilute, till after three or four revolutions they ended in perfect whiteness. Their form, when the Glasses were most compress'd so as to make the black Spot appear in the center, is delineated in the [thirty-first] Figure; where *a, b, c, d, e: f, g, h, i, k: l, m, n, o, p: q, r: s, t: v, x: y, z,* denote the Colours reckon'd in order from the center, black, blue, white, yellow, red: violet, blue, green, yellow, red: purple, blue, green, yellow, red: green, red: greenish blue, red: greenish blue, pale red: greenish blue, reddish white.

By "circuit" Newton means a cycle that includes all or some of the spectral colors in spectral sequence (possibly also white), without repetition, beginning with violet (or reddish violet, i.e., purple); where purple or violet does not occur, a new cycle begins with blue or, when no blue appears, green.

Obs. 5. To determine the interval of the Glasses, or thickness of the interjacent Air, by which each Colour was produced, I measured the Diameters of the first six Rings at the most lucid part of their Orbits, and squaring them, I found their Squares to be in the arithmetical Progression of the odd Numbers, 1, 3, 5, 7, 9, 11. And since one of these Glasses was plane, and the other spherical, their Intervals at those Rings must be in the same Progression. I measured also the Diameters of the dark or faint Rings between the more lucid Colours, and found their Squares to be in the arithmetical Progression of the even Numbers, 2, 4, 6, 8, 10, 12. And it being very nice and difficult to take these measures exactly; I repeated them divers times at divers parts of the Glasses, that by their Agreement I might be confirmed in them. . . .

The question Newton addresses now is how these colors correlate with the thickness of the layer of air between the lenses. He measures the distance between the brightest, most lucid hue of each cycle and the central black spot (presumably with a geometric compass), multiplies this number by itself, and finds the indicated extended ratios (see Appendix B). This calculation yields a series that has the same proportions as the odd numbers, whereas the darkest hues of each cycle give the ratios of the successive even numbers. This is a very curious, very tantalizing mathematical regularity, leading the mathematically inclined Newton to suspect he is on the track of something important.

Obs. 6. The Diameter of the sixth Ring at the most lucid part of its Orbit was 58/100 parts of an Inch, and the Diameter of the Sphere on which the double convex Object-glass was ground was about 102 Feet, and hence I gathered the thickness of the Air or Aereal Interval of the Glasses at that Ring. But some time after, suspecting that in making this Observation I had not determined the Diameter of the Sphere [on which the lens had been ground] with sufficient accurateness, and being uncertain whether the Plano-convex Glass was truly plane, and not something concave or convex on that side which I accounted plane; and whether I had not pressed the Glasses together, as I often did, to make them touch; (For by pressing such Glasses together their parts easily yield inwards, and the Rings thereby become sensibly broader than they would be, did the Glasses keep their Figures.) I repeated the Experiment, and found the Diameter of the sixth lucid Ring about 55/100 parts of an Inch. I repeated the Experiment also with such an Object-glass of another Telescope as I had at hand. This was a double Convex ground on both sides to one and the same Sphere, and its Focus was distant from it 83 2/5 Inches. And thence, if the Sines of Incidence and Refraction of the bright yellow Light be assumed in proportion as 11 to 17, the Diameter of the Sphere to which the Glass was figured will by computation be found 182 Inches.[4] This Glass I laid upon a flat one, so that the black Spot appeared in the middle of the Rings of Colours without any other Pressure than that of the weight of the Glass. And now measuring the Diameter of the fifth dark Circle as accurately as I could, I found it the fifth part of an Inch precisely. This Measure was taken with the points of a pair of Compasses on the upper Surface on the upper Glass, and my Eye was about eight or nine Inches distance from the Glass, almost perpendicularly over it, and the Glass was $1/6$ of an Inch thick, and thence it is easy to collect that the true Diameter of the Ring between the Glasses was greater than its measur'd Diameter above the Glasses in the Proportion of 80 to 79, or thereabouts, and by consequence equal to $16/79$ parts of an Inch, and its true Semi-diameter equal to $8/79$ parts. Now as the Diameter of the Sphere (182 Inches) is to the Semi-diameter of this fifth dark Ring ($8/79$ parts of an Inch) so is this Semi-diameter to the thickness of the Air at this fifth dark Ring; which is therefore $32/567,931$ or $100/1,774,784$ Parts of an Inch; and the fifth Part thereof, *viz.* the $1/88,739$ Part of an Inch, is the Thickness of the Air at the first of these dark Rings.

Here we see at work Newton's technique of determining the "aereal interval" by making use of his knowledge of the curvature of the lenses and by correcting for the fact that the refraction of the ray as it enters and emerges from the glass results in a change in its direction that creates a discrepancy between the ray's apparent place of origin in the plate of air and its actual origin. The question is not simply where the ring *appears* to be but where the rays that produce the ring *actually undergo* refraction and reflection; just as with a submerged oar, the apparent location combined with the laws of optics allows us to determine the actual location. In this case, Newton calculates that the actual diameter of the relevant "aereal" ring is slightly larger than the ring measured at the upper surface of the biconvex lens. By using geometrical and trig-onometric relationships and also exploiting his earlier experimental de-termination that the thicknesses of the air films for the first five dark rings are in the proportions of the even numbers 2, 4, 6, 8, 10, he is able to calculate the thickness of the first dark ring.

In the next observation Newton continues to expand his acquaintance with the conditions governing these phenomena by doing similar meas-urements and calculations for different positions of the eye, in order to establish the proportions that hold between these factors. He eventually arrives at a rule:

... And from these Measures I seem to gather this Rule: That the Thickness of the Air is proportional to the Secant of an Angle, whose Sine is a certain mean Proportional between the Sines of Incidence and Refraction. And that mean Proportional, so far as by these Measures I can determine it, is the first of an hundred and six arithmetical mean Proportionals between those Sines counted from the bigger Sine, that is, from the Sine of Refraction when the Refraction is made out of the Glass into the Plate of Air, or from the Sine of Incidence when the Refraction is made out of the Plate of Air into the Glass.[5]

This is an empirical rule or law, that is, a rule or law derived from experimental observation, which relates the incidence and refraction of the ray passing through the plate of air to the thickness of that plate of air. Newton gives no account of how precisely he arrived at this particular formula; it has the earmarks of being a rule of thumb or educated guess reached by persistent attempts at calculation and approximation.

In the ninth observation Newton observes the rings formed by light that is transmitted through, rather than reflected by, the two lenses; he finds that the colors are what we today would call the complementaries of those observed in reflected light. Next he wets the lenses so that water rather than air will form the "interjacent" medium, thereby reducing the diameter and vividness of the rings. Then he turns to using prismatic colors as the illuminating light, that is, homogeneal or monochromatic light, in a camera obscura (the previous experiments were performed "in the open Air"), which enables him to see more rings (up to twenty). More importantly for his theory, he finds "the Circles which the red Light made to be manifestly bigger than those which were made by the blue and violet." Once more he is eager to offer a musical analogy (in Observation 14):

... And hence I seem to collect that the thicknesses of the Air between the Glasses there, where the Ring is successively made by the limits of the five principal Colours (red, yellow, green, blue, violet) in order (that is, by the extreme red, by the limit of red and yellow in the middle of the orange, by the limit of yellow and green, by the limit of green and blue, by the limit of blue and violet in the middle of the indigo, and by the extreme violet) are to one another very nearly as the sixth lengths of a Chord which found the Notes in a sixth Major, *sol, la, mi, fa, sol, la.* But it agrees something better with the Observations to say, that the thicknesses of the Air between the Glasses there, where the Rings are successively made by the limits of the seven Colours, red, orange, yellow, green, blue, indigo, violet in order, are to one another as the Cube Roots of the Squares of the eight lengths of a Chord, which found the Notes in an eighth, *sol, la, fa, sol, la, mi, fa, sol;* that is, as the Cube Roots of the Squares of the Numbers, 1, 8/9, 5/6, 3/4, 2/3, 3/5, 9/16, 1/2.

The use of the cube roots of squares is probably in part an ad hoc attempt to provide a good fit between data and hypothesis rather than a deduction from mathematical principles, but it also establishes a relationship between the rings and the division of the spectrum according to the musical scale analogy. (Note that Newton says that there is a better fit if one takes seven colors instead of five, although both orange and indigo are even less conspicuous with thin-film colors than they are with the prismatic spectrum.) If this formula were strictly true, it would deepen the meaning of cosmic harmony and perhaps also establish a concrete analogy between the appearances of color and certain astronomical regularities, namely, the proportions discovered by Kepler that hold between the periods of revolution of the planets and their distances from the sun: the square of their period is in a constant proportion to the cube of their distance. (In the *Principia* Newton showed that this can be mathematically deduced from the inverse square law of universal gravitation.) The circles of colors would thus bear a certain analogy to the circles of the planets.

Newton goes on to note that not only the circles formed in *reflected* light but also those formed in *transmitted* light are of the spectral color used as illuminant, and the sequence of the dark and the colored circles is governed by the same arithmetical progression of the squares of even and odd numbers encountered in the fifth observation. After observing the thin-film colors exhibited by soap bubbles, Newton makes several concluding observations, one using thin sheets of mica or Muscovy glass (to show that not the material but only its thickness is responsible for the colors seen), another using a prism to observe the rings formed in white light (the effect is to make the rings more distinct and numerous on the side of the dark central spot closer to the prism while washing out the rings on the opposite side). He also draws two conclusions: that a thin plate made of a dense material in a rarer medium (for example, a thin glass lens in air) produces more vivid colors than a rare material in a denser medium (air between two glasses), and that the inmost rings reflect a greater quantity of light than the ones farther out.

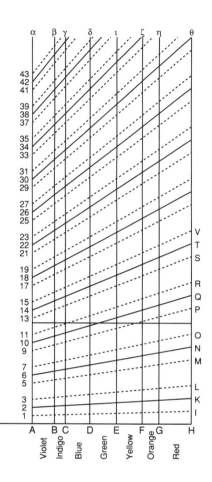

Fig. 32

Newton presents his theory of the cause of these rings in the third part of Book II. But before that, in the second part, he makes "Remarks upon the foregoing Observations," which are themselves already theoretical reflections on their significance. "Having given my Observations of these Colours, before I make use of them to unfold the Causes of the Colours of natural Bodies, it is convenient that by the simplest of them . . . I first explain the more compounded." He begins by creating a schematic diagram (Fig. 32). A line segment is divided into seven parts by eight points, whose positions correspond to the proportions he remarked in Observation 14, the cube roots of the squares of numbers expressing the proportions of the spectral colors conceived according to the musical scale, 1, 8⁄9, 5⁄6, 3⁄4, 2⁄3, 3⁄5, 9⁄16, 1⁄2.

Newton calculated the values for this sequence of cube roots of squares as 0.6300, 0.6814, 0.7114, 0.7631, 0.8255, 0.8855, 0.9243, and 1.0000. (These are all correct to four significant digits, except for the seventh, which should be 0.9245.) In Fig. 32, if the segment YH is taken to be 1 unit long, then YG is 0.9243, YF 0.8855, YE 0.8255, YD 0.7631, YC 0.7114, YB 0.6814, YA 0.6300. Notice that, in contrast to the arrangement of the colors according to the ratios of the musical scale in Part II of Book I, the lengths assigned to the colors on the red side of the diagram are larger than those on the violet side. From these eight points A through H he erects perpendiculars to the line YH. Any height along these perpendiculars is intended to represent the thickness of a thin film. The first perpendicular Aα is divided into equal units, numbered by consecutive numbers 1, 2, 3, and so forth, but leaving every fourth point unnumbered (thus the sequence goes 1, 2, 3, skip the next point of division, 5, 6, 7, skip a point, 9, 10, 11, skip, and so on). Finally, Newton draws lines from Y through each of the points 1, 2, 3; 5, 6, 7; 9, 10, 11; and so on, although he suppresses from the figure the portions of these lines that would be to the left of perpendicular Aα.

What is the logic of this diagram? The segment AB represents the violet spectral colors, BC the indigo, CD the blue, and so forth, to GH for red. On the vertical line Aα the length A2 indicates the relative thickness of the thin plate at which the extreme violet of the first colored cycle or ring is most abundantly reflected, A6 the thickness at which the violet of the second colored ring is most reflected, A10 the thickness for violet's abundant reflection in the third ring, and so on. Note that the relative thicknesses are 2 units, 6, 10, 14, 18, and so forth; that is, they are in the proportion 1, 3, 5, 7, 9, and so forth, in accordance with what was found out in earlier observations. Similarly, the length from B to the intersection of line 2K with Bβ represents the thickness at which the color between violet and indigo is most copiously reflected in the first cycle of rings, the length from C to the intersection of 2K with Cγ the thickness for indigo-blue, and so on to HK, which stands for the thickness at which extreme red is most copiously reflected. The lengths from B to the intersection of 6N with Bβ, from B to the intersection of 10Q with Bβ, and so on, are the thicknesses for the copious reflection of violet-indigo in the second, third, and subsequent cycles, and similarly for the other colors along their respective perpendiculars.

Since all rings visible in homogeneal light have breadth, it cannot be the case that extreme spectral violet is reflected only at thickness A2, for then the first violet circle (and all other circles formed by spectral light) would be as narrow as a point. So Newton represents a range of thicknesses at which at least some of the extreme violet will be reflected, from A1 to A3 in the first cycle, from A5 to A7 in the second, and so on, with a similar conclusion for the other colors. (By the way, between the lines 3L and 5M all the spectral lights are transmitted rather than reflected.)

Now we can understand the use of this diagram. Suppose someone asks: do we have some way of knowing what rays make up the light that is reflected to our eyes from a particular ring? The answer is derived by taking a ruler and laying it parallel to the base line YAH at the point representing the thickness of the thin film at which that ring is seen. If in the diagram we laid the ruler down a little above point 1, representing 1 unit of thickness, we would note that a thin film of this thickness should reflect all types of light, since the ruler line falls entirely above the line 1I (so we would probably see white there). If we put the ruler precisely at 1, only a small amount of extreme violet would be reflected, so the inmost fringe of the first cycle would be violet. The line drawn from just below 13 to just above O shows that as the thickness increases we get more interesting results. There (slightly less than 13 units of thickness), there will be no violet or indigo reflected, nor any orange or red; only blue, yellow, and, preeminently, green of the third cycle will be reflected, and so a shade of green should be visible. As you place the ruler farther up the diagram you will find that a greater variety of spectral lights from *different* cycles will be combined in the reflected light, and so it is increasingly likely that the result will be white.

From this diagram, combined with a table indicating the absolute thicknesses of air, water, and glass that produce the various colors, plus an earlier table that described the effect on the position of the circles of oblique viewing angles, Newton explains the appearances as the overlapping of the "component" circles produced at each point by the different kinds of homogeneal light. "Now as all these things follow from the properties of Light by a mathematical way of reasoning, so the truth of them may be manifested by Experiments." He concludes by showing that the observation of the thin-film colors through a prism produces phenomena that confirm the results of Book I, that is, differential refraction and the theory of colors. In particular, it is confirmed that white is a mixture of all kinds of rays and that there is a constant relationship between color and refrangibility.

> . . . Whence it follows, that the colorifick Dispositions of Rays are also connate [inborn] with them, and immutable; and by consequence, that all the Productions and Appearances of Colours in the World are derived, not from any physical Change caused in Light by Refraction or Reflexion, but only from the various Mixtures or Separations of Rays, by virtue of their different Refrangibility or Reflexibility. And in this respect the Science of Colours becomes a Speculation as truly mathematical as any other part of Opticks. I mean, so far as they depend on the Nature of Light, and are not produced or alter'd by the Power of Imagination, or by striking or pressing the Eye.

The statement here echoes one he had written thirty years earlier in his February 1672 letter, in a passage that was suppressed from the printed version. The theory of differential refrangibility according to color mathematized the science of light and colors in an unprecedented way, without Newton's having to pronounce on the fundamental nature of light or to advance an artificial hypothesis about the mechanisms involved. Likewise here for thin-plate colors. But Newton is not yet done, for in the concluding parts of Book II he will lay the foundation for a theory of the interaction of light and matter that purports to explain exhaustively the physical causes of the phenomena of light and colors and that turns the appearance of color into an index for the internal constitution of matter.

Chapter 8

The *Opticks,* Book II, Parts III and IV

In the previous two parts of Book II Newton has shown not only that thin-film colors can be measured, quantified, and predicted but also that they are somehow connected to the reflection and transmission of light rays at interfaces between different transparent media. This is the reason for Newton's mentioning the property of reflexibility at the end of Part II: the reflection of light, which had hitherto been more or less taken for granted, becomes something of a puzzle (for which Newton now appears to have at least the beginnings of an answer). At interfaces between two media some light is reflected and some not. Why? We tend to think of the colors of bodies as the result of light "bouncing off" (being reflected from) them. But how does this really happen, and what underlies it? The property of the reflexibility of rays is what is at issue in these questions.[1]

There is also a quite remarkable achievement that will be generalized and extended here. Think back, first, to Book I: from the intimate connection between refrangibility and color there was developed a science of isolating and measuring rays of light. We could call that an achievement internal to optics. In the first two parts of Book II, Newton shows that there is an intimate connection between the dimensions of thin films and the rays that are reflected or transmitted, and he is able to set up diagrams, tables, and formulas that allow us to correlate both relative and absolute measures of the dimensions of thin films with the colors we see. That is, Newton shows how it might be possible to measure very small dimensions (on the order of $\frac{1}{100,000}$ of an inch) of transparent materials by means of the theory of colored rings. We can call this a result that, although discovered within optics, provides a tool that can be used in other scientific fields to explore matter at a dimension much smaller than anything that had previously been possible. Thus the properties of light can be used to investigate the inner constitution of things; light is no longer just an object of study in its own right but even more an instrument, a key, to be used in the exploration and quantitative determination of every material thing.

THE SECOND BOOK OF OPTICKS.

PART III.

Of the permanent Colours of natural Bodies, and the Analogy between them and the Colours of thin transparent Plates.

I am now come to another part of this Design, which is to consider how the Phaenomena of thin transparent Plates stand related to those of all other natural Bodies. Of these Bodies I have already told you that they appear of divers Colours, accordingly as they are disposed to reflect most copiously the Rays originally endued with those Colours. But their Constitutions, whereby they reflect some Rays more copiously than others, remain to be discover'd; and these I shall endeavor to manifest in the following Propositions.

Newton's way is no longer either experimentally demonstrative or inductive; rather, he intends to establish fundamental principles derived from the experiments of Part I, from the explanatory tools provided in Part II, and from other phenomena he will bring up as he goes along. The first principle he lays down is that materials that refract most strongly also reflect most strongly.

PROP. I. *Those Superficies of transparent Bodies reflect the greatest quantity of Light, which have the greatest refracting Power; that is, which intercede Mediums that differ most in their refractive Densities. And in the Confines of equally refracting Mediums there is no Reflexion.*

The Analogy between Reflexion and Refraction will appear by considering, that when Light passeth obliquely out of one Medium into another which refracts from the perpendicular, the greater is the difference of their refractive Density, the less Obliquity of Incidence is requisite to cause a total Reflexion. For as the Sines are which measure the Refraction, so is the Sine of Incidence at which the total Reflexion begins, to the Radius of the Circle; and consequently that Angle of Incidence is least where there is the greatest difference of the Sines. Thus in the passing of Light out of Water into Air, where the Refraction is measured by the Ratio of the Sines 3 to 4, the total Reflexion begins when the Angle of Incidence is about 48 Degrees 35 Minutes. In passing out of Glass into Air, where the Refraction is measured by the Ratio of the Sines 20 to 31, the total Reflexion begins when the Angle of Incidence is 40 Degrees 10 Minutes; and so in passing out of Crystal, or more strongly refracting Mediums into Air, there is still a less obliquity requisite to cause a total reflexion. Superficies therefore which refract most do soonest reflect all the Light which is incident on them, and so must be allowed most strongly reflexive.

Newton is making a correlation between refraction and total internal reflection. Once they are inside a very dense material, only rays that strike nearly perpendicularly to an interface can escape; all the rest are totally reflected. How this helps explain reflection in general begins to appear in the next propositions, which present a theory of matter as consisting of small parts that are more or less transparent.

. . . PROP. II. *The least parts of almost all natural Bodies are in some measure transparent: And the Opacity of those Bodies ariseth from the multitude of Reflexions caused in their internal Parts.*

That this is so has been observed by others, and will easily be granted by them that have been conversant with Microscopes. And it may be also tried by applying any substance to a hole through which some Light is immitted [sent] into a dark Room. For how opake soever that Substance may seem in the open Air, it will by that means appear very manifestly transparent, if it be of a sufficient thinness. Only white metalline Bodies must be excepted, which by reason of their excessive density seem to reflect almost all the Light incident on their first Superficies; unless by solution in Menstruums [solvents] they be reduced into very small Particles, and then they become transparent.

Newton now combines these first two propositions to argue that if these smallest parts were not separated by either empty space or matter of a different (especially lesser) density, bodies would not reflect much light but rather would absorb it or let it pass through.

PROP. III. *Between the parts of opake and colour'd Bodies are many Spaces, either empty, or replenish'd with Mediums of other Densities; as Water between the tinging Corpuscles wherewith any Liquor is impregnated, Air between the aqueous Globules that constitute Clouds or Mists; and for the most part Spaces void of both Air and Water, but yet perhaps not wholly void of all Substance, between the parts of hard Bodies.*

The truth of this is evinced by the two precedent Propositions: For by the second Proposition there are many Reflexions made by the internal parts of Bodies, which, by the first Proposition, would not happen if the parts of those Bodies were continued without any such Interstices between them; because Reflexions are caused only in Superficies, which intercede Mediums of a differing density, by *Prop.* 1.

But farther, that this discontinuity of parts is the principal Cause of the opacity of Bodies, will appear by considering, that opake Substances become transparent by filling their Pores with any Substance of equal or almost equal density with their parts. Thus Paper dipped in Water or Oil . . . and many other Substances soaked in such Liquors as will intimately pervade their little Pores, become by that means more transparent than otherwise; so, on the contrary, the most transparent Substances may, by evacuating their Pores, or separating their parts, be render'd sufficiently opake; as . . . Glass by being reduced to Powder, . . . and Water by being form'd into many small Bubbles. . . .

So if we could examine an opaque (i.e., reflecting) body at very great magnification we would find that its microstructure consists of transparent or semitransparent chips and pieces separated by spaces that may be filled by other matter. How large are these pieces, and how far apart?

PROP. IV. *The Parts of Bodies and their Interstices must not be less than of some definite bigness, to render them opake and colour'd.*

For the opakest Bodies, if their parts be subtilly divided, (as Metals, by being dissolved in acid Menstruums, &c.) become perfectly transparent. And you may also remember, that in the eighth Observation there was no sensible Reflexion at the Superficies of the Object-glasses, where they were very near one another, though they did not absolutely touch. And in the 17th Observation the Reflexion of the Water-bubble where it became thinnest was almost insensible, so as to cause very black Spots to appear on the top of the Bubble, by the want of reflected Light.

On these grounds I perceive it is that Water, Salt, Glass, Stones, and such like Substances, are transparent. For, upon divers Considerations, they seem to be as full of Pores or Interstices between their parts as other Bodies are, but yet their Parts and Interstices to be too small to cause Reflexions in their common Surfaces.

Newton now has laid down the basic elements of a theory of bodies. Despite their appearance of being solid, they are full of spaces or pores, which in some bodies are very large. The pores can be occupied by matter (fluids) of greater or lesser density than the solid particles, and this affects the transparency or opaqueness of the macroscopic body, because whether the small particles (which are like thin plates) transmit or reflect light depends on both their thickness and the difference of their density from the density of the surrounding matter. If the solid particles are thin enough, they will transmit all light (so the macroscopic body will be transparent), as will thicker particles surrounded by matter of the same density. Opaque bodies consist of small particles, too, but they must be of an adequate thickness, and any matter pervading the pores of the body must be of considerably different density.[2]

Having established that the theory of thin plates gives reasons to think that there are important microscopic differences between transparent and opaque bodies, Newton now turns to the question of why opaque bodies have different colors. He starts by providing an easy way of conceiving macroscopic bodies as made up of thin plates.

PROP. V. *The transparent parts of Bodies, according to their several sizes, reflect Rays of one Colour, and transmit those of another, on the same grounds that thin Plates or Bubbles do reflect or transmit those Rays. And this I take to be the ground of all their Colours.*

For if a thinn'd or plated Body, which being of an even thickness, appears all over of one uniform Colour, should be slit into Threads, or broken into Fragments, of the same thickness with the Plate; I see no reason why every Thread or Fragment should not keep its Colour, and by consequence why a heap of those Threads or Fragments should not constitute a Mass or Powder of the same Colour, which the Plate exhibited before it was broken. And the parts of all natural Bodies being like so many Fragments of a Plate, must on the same grounds exhibit the same Colours.

That is, imagine that a uniform, thin solid were divided into a large number of small pieces having the same thinness as the whole solid: this should not at all change the thin-plate colors one sees. Similarly, one can imagine these small pieces jumbled together: they should still produce the same kinds of thin-plate colors. But this would not be much different from the state of ordinary bodies, which at microscopic dimensions are probably a jumble of small but more or less equal particles. Newton is leading his reader to apply imaginatively, step by step, the lessons of the investigation.

In the next proposition, Newton directly applies results about thin-plate colors from Parts I and II of Book II. First, the theory he is developing will work best if the particles are much denser than the matter or medium pervading the pores.

... PROP. VI. *The parts of Bodies on which their Colours depend, are denser than the Medium which pervades their Interstices.*

This will appear by considering, that the Colour of a Body depends not only on the Rays which are incident perpendicularly on its parts, but on those also which are incident at all other Angles. And that according to the 7th Observation, a very little variation of obliquity will change the reflected Colour, where the thin Body or small Particle is rarer than the ambient Medium, insomuch that such a small Particle will at diversly oblique Incidences reflect all sorts of Colours, in so great a variety that the Colour resulting from them all, confusedly reflected from a heap of such Particles, must rather be a white or grey than any other Colour, or at best it must be but a very imperfect and dirty Colour. Whereas if the thin Body or small Particle be much denser than the ambient Medium, the Colours, according to the 19th Observation, are so little changed by the variation of obliquity, that the Rays which are reflected least obliquely may predominate over the rest, so much as to cause a heap of such Particles to appear very intensely of their Colour.

Newton has forestalled here the objection that, since thin-plate colors depend on the angle at which you view them, any body colors produced by this phenomenon would constantly shimmer and change as the viewing position changed (of course there are some objects like this, for instance the feathers of a pigeon and fish scales). If the small particles are plates of *similar* dimensions (as suggested in Proposition V), and if they are quite *dense* (meaning that obliquity will have a lesser effect on the appearance of colors), then the observations of thin plates in the preceding parts of Book II suggest that the color will not markedly change with the viewing angle.

Body colors, it seems, depend vitally on the microstructure of bodies. In Part II Newton showed that we could establish correlations between the colors that appear in thin plates and the dimensions of those plates. Now he extends this discovery to bodies in general: the colors they display reveal something about the dimensions of their smallest parts.

... PROP. VII. *The bigness of the component parts of natural Bodies may be conjectured by their Colours.*

For since the parts of these Bodies, by *Prop.* 5. do most probably exhibit the same Colours with a Plate of equal thickness, provided they have the same refractive density; and since their parts seem for the most part to have much the same density with Water or Glass, as by many circumstances is obvious to collect; to determine the sizes of those parts, you need only have recourse to the . . . Tables [not reproduced here], in which the thickness of Water or Glass exhibiting any Colour is expressed. . . .

The greatest difficulty is here to know of what Order the Colour of any Body is. . . .

If we recall that the colored rings are generated in cycles of bright and dark colors, and that a given color may appear in different cycles (*orders,* as Newton now calls them), we recognize that to determine the thickness of the film of air that has given rise to a particular hue we must first ascertain which cycle or order it comes from. The specific qualities of hue, according to Newton, provide the chief clue. He proceeds to discuss the various colors that one sees in bodies and relates them to the color seen in the different cycles of thin-plate colors. Moreover, he tries to incorporate some of his chemical knowledge by correlating changes in body colors with effects by chemical processes on the size and density of particles.

Scarlets, and other *reds, oranges,* and *yellows,* if they be pure and intense, are most probably of the second order. Those of the first and third order also may be pretty good; only the yellow of the first order is faint, and the orange and red of the third Order have a great Mixture of violet and blue.

There may be good *Greens* of the fourth Order, but the purest are of the third. And of this Order the green of all Vegetables seems to be, partly by reason of the Intenseness of their Colours, and partly because when they wither some of them turn to a greenish yellow, and others to a more perfect yellow or orange, or perhaps to red, passing first through all the aforesaid intermediate Colours. Which Changes seem to be effected by the exhaling of the Moisture which may leave the tinging Corpuscles more dense, and something augmented by the Accretion of the oily and earthy Part of that Moisture. Now the green, without doubt, is of the same Order with those Colours into which it changeth, because the Changes are gradual, and those Colours, though usually not very full, yet are often too full and lively to be of the fourth Order.

Blues and *Purples* may be either of the second or third Order, but the best are of the third. Thus the Colour of Violets seems to be of that Order, because their

Syrup by acid Liquors turns red, and by urinous and alcalizate turns green. For since it is of the Nature of Acids to dissolve or attenuate, and of Alcalies [alkalis] to precipitate or incrassate [make denser], if the Purple Colour of the Syrup was of the second Order, an acid Liquor by attenuating its tinging Corpuscles would change it to a red of the first Order, and an Alcali by incrassating them would change it to a green of the second Order; which red and green, especially the green, seem too imperfect to be the Colours produced by these Changes. But if the said Purple be supposed of the third Order, its Change to red of the second, and green of the third, may without any Inconvenience be allow'd.

. . . .

The *blue* of the first Order, though very faint and little, may possibly be the Colour of some Substances; and particularly the azure Colour of the Skies seems to be of this Order. For all Vapours when they begin to condense and coalesce into small Parcels, become first of that Bigness, whereby such an Azure must be reflected before they can constitute Clouds of other Colours. And so this being the first Colour which Vapours begin to reflect, it ought to be the Colour of the finest and most transparent Skies, in which Vapours are not arrived to that Grossness requisite to reflect other Colours, as we find it is by Experience.

Whiteness, if most intense and luminous, is that of the first Order, if less strong and luminous, a Mixture of the Colours of several Orders. . . .

Lastly, for the production of *black,* the Corpuscles must be less than any of those which exhibit Colours. For at all greater sizes there is too much Light reflected to constitute this Colour. But if they be supposed a little less than is requisite to reflect the white and very faint blue of the first order, they will . . . reflect so very little Light as to appear intensly [sic] black, and yet may perhaps variously refract it to and fro within themselves so long, until it happen to be stifled and lost, by which means they will appear black in all positions of the Eye without any transparency. And from hence may be understood why Fire, and the more subtle dissolver Putrefaction, by dividing the Particles of Substances, turn them to black, why small quantities of black Substances impart their Colour very freely and intensely to other Substances to which they are applied; the minute Particles of these, by reason of their very great number, easily overspreading the gross Particles of others. . . .

In these Descriptions I have been the more particular, because it is not impossible but that Microscopes may at length be improved to the discovery of the Particles of Bodies on which their Colours depend, if they are not already in some measure arrived to that degree of perfection. For if those Instruments are or can be so far improved as with sufficient distinctness to represent Objects five or six hundred times bigger than at a Foot distance they appear to our naked Eyes, I should hope that we might be able to discover some of the greatest of those Corpuscles. And by one that would magnify three or four thousand times perhaps they might all be discover'd, but those which produce blackness. In the mean while I see nothing material in this Discourse that may rationally be doubted of, excepting this Position: That transparent Corpuscles of the same thickness and density with a Plate, do exhibit the same Colour.[3] And this I would have understood not without some Latitude, as well because those Corpuscles may be of irregular Figures, and many Rays must be obliquely incident on them, and so have a shorter way through them than the length of their Diameters, as because the straitness [being confined] of the Medium put in on all sides within such Corpuscles may a little alter its Motions or other qualities on which the Reflexion depends. But yet I cannot much suspect the last, because I have observed of some small Plates of Muscovy Glass [mica] which were of an even thickness, that through a Microscope they have appeared of the same Colour at their edges and corners where the included Medium was terminated, which they appeared of in other places. However it will add much to our Satisfaction, if those Corpuscles can be discover'd with Microscopes; which if we shall at length attain to, I fear it will be the utmost improvement of this Sense. For it seems impossible to see the more secret and noble Works of Nature within the Corpuscles by reason of their transparency.

Although Newton does not consider these results to be as certain and exact as those of Book I—and some of them are quite speculative, especially those that make conjectures about chemical transformations— he nevertheless has confidence that thin-plate colors open up a program of research that will lead to the discovery of all but perhaps the smallest of body particles. What goes on within such particles, however, may remain forever hidden, because it will not be possible to see them by means of light.

Although Newton was wrong about the physical mechanisms of reflection and by many orders of magnitude about the precise limits of optical magnification, and although in the past century physicists have developed more powerful techniques than he could have hoped for, using x-rays, electrons, and other kinds of radiation and particles, this does not change the continuing relevance of Newton's assertion that beyond some level of the very small the truth of things remains hidden from us; at the very least, we will never be able to see the ultimate arrangement of matter directly.

PROP. VIII. *The Cause of Reflexion is not the impinging of Light on the solid or impervious parts of Bodies, as is commonly believed.*

This will appear by the following Considerations. First, That in the passage of Light out of Glass into Air there is a Reflexion as strong as in its passage out of Air into Glass, or rather a little stronger, and by many degrees stronger than in its passage out of Glass into Water. And it seems not probable that Air should have more strongly reflecting parts than Water or Glass. . . . Secondly, If Light in its passage out of Glass into Air be incident more obliquely than at an Angle of 40 or 41 Degrees it is wholly reflected, if less obliquely it is in great measure transmitted. Now it is not to be imagined that Light at one degree of obliquity should meet with Pores enough in the Air to transmit the greater part of it, and at another degree of obliquity should meet with nothing but parts to reflect it wholly, especially considering that in its passage out of Air into Glass, how oblique soever be its Incidence, it finds Pores enough in the Glass to transmit a great part of it. . . . Thirdly, If the Colours made by a Prism placed at the entrance of a Beam of Light into a darken'd Room be successively cast on a second Prism placed at a greater distance from the former, in such manner that they are all alike incident upon it, the second Prism may be so inclined to the incident Rays, that those which are of a blue Colour shall be all reflected by it, and yet those of a red Colour pretty copiously transmitted. Now if the Reflexion be caused by the parts of Air or Glass, I would ask, why at the same Obliquity of Incidence the blue should wholly impinge on those parts, so as to be all reflected, and yet the red find Pores enough to be in a great measure transmitted. Fourthly, Where two Glasses touch one another, there is no sensible Reflexion, as was declared in the first Observation; and yet I see no reason why the Rays should not impinge on the parts of Glass, as much when contiguous to other Glass as when contiguous to Air. Fifthly, When the top of a Water-Bubble . . . by the continual subsiding and exhaling of the Water grew very thin, there was such a little and almost insensible quantity of Light reflected from it, that it appeared intensely black; whereas round about that black Spot, where the Water was thicker, the Reflexion was so strong as to make the Water seem very white. . . .Sixthly, If Reflexion were caused by the parts of reflecting Bodies, it would be impossible for thin Plates or Bubbles, at one and the same place, to reflect the Rays of one Colour, and transmit those of another. . . . Lastly, Were the Rays of Light reflected by impinging on the solid parts of Bodies, their Reflexions from polish'd Bodies could not be so regular as they are. For in polishing Glass with Sand, Putty, or Tripoly [a very fine polishing powder], it is not to be imagined that those Substances can, by grating and fretting the Glass, bring all its least Particles to an accurate Polish. . . . So then it remains a Problem, how Glass polish'd by fretting Substances can reflect Light so regularly as it does. And this Problem is scarce otherwise to be solved, than by saying, that the Reflexion of a Ray is effected, not by a single point of the reflecting Body, but by some power of the Body which is evenly diffused all over its Surface, and by which it acts upon the Ray without immediate Contact. For that the parts of Bodies do act upon Light at a distance shall be shewn hereafter.

Traditional theory of mechanical matter and motions conceived of small, massive particles that changed direction because of their direct contact with one another, in particular because of collision.[4] But if matter is as Newton has been suggesting, the fundamental optical phenomena that turn rays of light out of their paths cannot be due to collision or contact. What the last part of this proposition presents is the idea of a field of force: perhaps all interfaces consist not so much of solid matter as a sudden change in a force field, which gives rise to more uniform effects than could be derived from a mass of particles. This claim strikes a bridge between the *Opticks* and the physical principles of the *Principia,* which understands motion as governed by the inertia of matter and the attraction of all particles of matter for one another.

In the next proposition Newton states explicitly the close link in practice between refrangibility and reflexibility. At surfaces, if a ray is not reflected it is refracted, and vice versa. Whatever is the cause of the one must in some way be the cause of the other as well.

. . . PROP. IX. *Bodies reflect and refract Light by one and the same power, variously exercised in various Circumstances.*

In the next proposition Newton shows that there is at least a rough correlation between the density of a body and its refractive power, where refractive power is now understood as a force and the ray understood as a particle subject to the force. By using the diagram of a ray refracted as it just grazes the surface of a transparent body (so that there is virtually no component of motion perpendicular to the surface in the incident ray; see Appendix C and Fig. 38) he shows that if the refracting force is exerted evenly in a small space near the refracting surface it can be measured by the squares of a very easily derived line, and that this measure is very nearly equal to the density of the body. This analysis is related to a similar one Newton did on purely theoretical grounds in the *Principia* (Proposition XCIV, Theorem XLVIII of Book I); although the analysis in the *Opticks* does not prove that such a region of force exists at interfaces, it establishes that such a force is consistent with other things known about light and matter.

... PROP. X. *If Light be swifter in Bodies than in Vacuo, in the proportion of the Sines which measure the Refraction of the Bodies, the Forces of the Bodies to reflect and refract Light, are very nearly proportional to the densities of the same Bodies; excepting that unctuous* [oily] *and sulphureous Bodies refract more than others of this same density.*

The analysis in Proposition X assumes that light moves faster in bodies than in empty space. Although that seems odd to us, who know that the speed of light in a vacuum is faster than in any substance, it in fact is a very plausible assumption if light is conceived as a tiny body hurtling through space; in particular, it explains why a ray of light is refracted *toward* the perpendicular when it enters a denser medium (namely, some force is applied perpendicularly at the interface; it speeds up the ray and thereby diverts it to a path closer to the perpendicular). At any rate, Newton shared this assumption with many of his predecessors and contemporaries.

... I have hitherto explain'd the power of Bodies to reflect and refract, and shew'd, that thin transparent Plates, Fibres, and Particles, do, according to their several thicknesses and densities, reflect several sorts of Rays, and thereby appear of several Colours; and by consequence that nothing more is requisite for producing all the Colours of natural Bodies, than the several sizes and densities of their transparent Particles. But whence it is that these Plates, Fibres, and Particles, do, according to their several thicknesses and densities, reflect several sorts of Rays, I have not yet explain'd. To give some insight into this matter, and make way for understanding the next part of this Book, I shall conclude this part with a few more Propositions. Those which preceded respect the nature of Bodies, these the nature of Light: For both must be understood, before the reason of their Actions upon one another can be known. And because the last Proposition depended upon the velocity of Light, I will begin with a Proposition of that kind.

This last paragraph of Proposition X marks the beginning of a new phase in the *Opticks*. The reader now has everything he or she needs to account for the colors of bodies in terms of the sizes and densities of their transparent particles. But there is not yet an answer to the question why rays are reflected or refracted by some intervals but not by others. To answer this, it is necessary to address more directly what the nature of light is and how it interacts with ordinary matter. As a prelude, in Proposition XI Newton asserts the finite velocity of light as established by Olaus Roemer's investigations of the periods of the eclipses of Jupiter's moons. In the nine propositions that follow Newton lays down his theory of the cause of thin-plate colors, the so-called "fits of easy transmission and reflection." Although it is highly speculative and perhaps not entirely consistent, the basic rationale is simple enough. In effect, these fits are properties that regularly and cyclically change in a way that determines whether rays are reflected or refracted at an interface.

. . . PROP. XII. *Every Ray of Light in its passage through any refracting Surface is put into a certain transient Constitution or State, which in the progress of the Ray returns at equal Intervals, and disposes the Ray at every return to be easily transmitted through the next refracting Surface, and between the returns to be easily reflected by it.*

This is manifest by the . . . Observations. For by those Observations it appears, that one and the same sort of Rays at equal Angles of Incidence on any thin transparent Plate, is alternately reflected and transmitted for many Successions accordingly as the thickness of the Plate increases in arithmetical Progression. . . . And this alternate Reflexion and Transmission, as I gather by the 24th Observation, continues for above an hundred vicissitudes [repetitions of the interval], and by the Observations in the next part of this Book, for many thousands, being propagated from one Surface of a Glass Plate to the other, though the thickness of the Plate be a quarter of an Inch or above: So that this alternation seems to be propagated from every refracting Surface to all distances without end or limitation. . . .

. . . .

What kind of action or disposition this is; Whether it consists in a circulating or a vibrating motion of the Ray, or of the Medium, or something else, I do not here enquire. Those that are averse from assenting to any new Discoveries, but such as they can explain by an Hypothesis, may for the present suppose, that as Stones by falling upon Water put the Water into an undulating Motion, and all Bodies by percussion excite vibrations in the Air; so the Rays of Light, by impinging on any refracting or reflecting Surface, excite vibrations in the refracting or reflecting Medium or Substance, and by exciting them agitate the solid parts of the refracting or reflecting Body, and by agitating them cause the Body to grow warm or hot; that the vibrations thus excited are propagated in the refracting or reflecting Medium or Substance, much after the manner that vibrations are propagated in the Air for causing Sound, and move faster than the Rays so as to overtake them; and that when any Ray is in that part of the vibration which conspires with its Motion, it easily breaks through a refracting Surface, but when it is in the contrary part of the vibration which impedes its Motion, it is easily reflected; and, by consequence, that every Ray is successively disposed to be easily reflected, or easily transmitted, by every vibration which overtakes it. But whether this Hypothesis be true or false I do not here consider. I content my self with the bare Discovery, that the Rays of Light are by some cause or other alternately disposed to be reflected or refracted for many vicissitudes.

DEFINITION. *The returns of the disposition of any Ray to be reflected I will call its* Fits of easy Reflexion, *and those of its disposition to be transmitted its* Fits of easy Transmission, *and the space it passes between every return and the next return, the* Interval of its Fits.

Refrangibility, reflexibility, and the property of producing color as defined earlier are all properties of rays. The thin-plate colors reveal yet another property, but Newton's discussion here shows his unwillingness to settle whether it is a property of rays, or of the medium, or of something else (whatever it might be). Thin-plate colors are unlike the previously analyzed phenomena precisely in that they do not reveal an *unchanging* property of rays; if they are produced by a property of the rays, that property undergoes constant cyclical change. It is "set" at the moment it enters a thin plate (that is, when it passes through a surface) and then varies in a way that allows the ray to pass through the second surface of the plate if the distance between the surfaces takes on certain values but not if the distance is a little more or less. The property seems really to belong to the whole interactive situation of the ray traversing the thin plate.

The hypothesis that Newton introduces in fact treats the fits as a kind of wave. The ray sets up a temporary vibratory motion in the medium, a motion that goes out in advance of the ray itself and so actually moves faster than the ray. Just as when a pebble breaks the surface of a pond the waves continue to be generated for some time thereafter, so the "breaking" of the interface surface by the ray generates new "fit-waves" after the ray has passed, so that the waves also trail behind the ray! When the ray and one of its fit-wave crests arrive at the opposite surface at the same moment, the ray passes through; when a trough arrives with the ray the ray is reflected back into the plate. How seriously we are supposed to take this hypothesis is open to question, since Newton appears to introduce it only for the sake of those who think explanation always requires the introduction of a hypothesis, and the whole notion resembles a theory of light-matter interaction that had been proposed by Robert Hooke, a person Newton could otherwise scarcely tolerate, in their correspondence of 1672.

The theory of fits of easy transmission and reflection is a cause for both admiration and consternation. Newton had already settled for a conception of light according to which rays are tiny but isolable objects that have determinate properties. Unlike refraction, however, thin-plate colors display transient and cyclical characteristics. The hypothesis of fits is an extraordinarily ingenious way of preserving the corpuscular ray theory while accounting for the new phenomenon (note as well that Newton suggests it might be connected with light's ability to heat matter). On the other hand, it looks like an ad hoc hypothesis incoherently tacked on to the theory in order to make it appear to have more explanatory power than it really does (especially since the mechanism by which these rays operate would require a great deal more knowledge about the microscopic structure of the rays and the medium, the principles of their interaction, and the kinds of forces and effects that are operative). The wave theory devised a century after Newton of course explained the thin-plate phenomena far more simply, as resulting from the interference of waves in a special medium called the ether. Whatever one's reaction to the fits, however, Newton was the first to specify clearly and exactly the periodic character of thin-plate colors and to face the need for a theory that would accommodate it. And, after all, he admitted that there was still a great deal to investigate here.

The next two propositions further explicate the basic principles of the fits.

PROP. XIII. *The reason why the Surfaces of all thick transparent Bodies reflect part of the Light incident on them, and refract the rest, is, that some Rays at their Incidence are in Fits of easy Reflexion, and others in Fits of easy Transmission.*

This may be gather'd from the 24th Observation, where the Light reflected by thin Plates of Air and Glass, which to the naked Eye appear'd evenly white all over the Plate, did through a Prism appear waved with many Successions of Light and Darkness made by alternate Fits of easy Reflexion and easy Transmission, the Prism severing and distinguishing the Waves of which the white reflected Light was composed, as was explain'd above.

And hence Light is in Fits of easy Reflexion and easy Transmission, before its Incidence on transparent Bodies. And probably it is put into such Fits at its first emission from luminous Bodies, and continues in them during all its progress. For these Fits are of a lasting nature, as will appear by the next part of this Book.

This proposition both applies the theory of fits to phenomena studied in Book I and also lays the groundwork for the observations of thick-plate colors in the last part of Book II. We learn from it that the fit-waves continue indefinitely (as Proposition XI had already suggested); indeed, Newton says it is likely that light is put into fits at its origin. If light is already in a fit when it is emitted by its source, it is likelier that fits are really properties of the rays rather than a property of the encounter of the ray with an interface.

Proposition XIV confirms Propositions I, VIII, and especially IX: the causes of reflection and refraction are intimately connected.

... PROP. XIV. *Those Surfaces of transparent Bodies, which if the Ray be in a Fit of Refraction do refract it most strongly, if the Ray be in a Fit of Reflexion do reflect it most easily.*

The final six propositions, about which we shall make no further comments, then lay down a basis for applying the quantitative formulas derived in Book II, Part II, to the theory of fits.

A Remark on Book II, Part IV

Proposition XIII noted that fits are of a lasting nature, probably from the moment of emission. Part IV accordingly undertakes a study of some phenomena very similar to the colored rings of Part I but caused now by glass lenses and water droplets, that is, by transparent "plates" considerably thicker than the narrow dimensions of the thin plates in the earlier parts. This part does not introduce any fundamentally new considerations, however, but rather confirms that the principle of fits can be used to explain the behavior of light in the case of thick plates as well as thin.

Chapter 9

The *Opticks,* Book III

Book III begins with yet another category of phenomena of light, one that like thin-plate colors was a recent addition to the field of optics, to which their discoverer, Francesco Maria Grimaldi (1618–1663), in his 1665 book *Physico-Mathesis de lumine,* had given the name by which it is known today, diffraction. Grimaldi had noticed that when a small opaque body is placed in the path of a narrow beam of sunlight admitted into a camera obscura, the shadow it casts is wider than it should be if light passed bodies in a straight line, and furthermore that the shadow is fringed by colored bands running parallel to the shadow's edge. To account for these and similar phenomena, Grimaldi proposed that light is a kind of fluid susceptible to wave-like motions.

Grimaldi's phenomena and theory were published at about the time Newton was beginning his optical experiments. Newton's first public treatment of them was contained in the "Hypothesis" he sent to the Royal Society in December 1675. Diffraction was a challenge to Newton on several counts. First, any theory of light and colors that pretended to be comprehensive would have to explain all known phenomena of light. Second, Grimaldi had established diffractive effects as a new class of phenomena, an apparently new property of light, and the *Opticks,* with its announced intention of proposing and proving the properties of light by reason and experiment, therefore had to examine diffraction at least as thoroughly as it had thin-plate colors. Third, if Grimaldi's theory were right, that light is a fluid displaying wave effects, then Newton's theory of light, which emphasizes its particulate and corpuscular nature, would be wrong.

Newton takes up the study of diffraction but changes the name; he calls this property of light *inflection.* This is a case where names are more than arbitrary labels applied to things. Already in Newton's day *refractio* was a long-established term that etymologically suggested a sharp break in the path of a ray when it passes into a refracting medium. *Diffractio* was a parallel coinage that suggested light was broken up or spread about. But this kind of notion Newton had long associated with modificationist theories of light. (Moreover, that colors are produced where light passes a physical boundary must have struck Newton as the kind of phenomenon that could give new life to the claim of some modificationists that a boundary was necessary to produce colors.) "Inflection," on the other hand, suggests that the ray is *bent;* as we shall see presently, Newton conceives of its cause as some short-range force at the boundary of the object strong enough to deflect the ray out of its original path. With this kind of explanation, Newton is explicitly striking a bridge between the experimental investigation of light's properties and the kind of force-particle physics he had undertaken in the *Principia*.

But this is not the only, nor the most important, sense in which the third book extends the project of the *Opticks*. The relatively brief and inconclusive study of inflection/diffraction points to the existence of properties and forces of matter that only future investigation would be able to penetrate. The queries that follow the diffraction experiments are bold speculations and speculative questions about the nature of light, colors, vision, the fundamental interactions of light with other matter, and the basic constitution of the universe.[1] Thus the *Opticks* won its reputation not just as a premier example of experimental science but also as a sketch of the entire physical cosmos as conceived by the mature mind of Isaac Newton.

THE THIRD BOOK OF OPTICKS.

PART I.

Observations concerning the Inflexions of the Rays of Light, and the Colours made thereby.

Grimaldo [Grimaldi] has inform'd us, that if a beam of the Sun's Light be let into a dark Room through a very small hole, the Shadows of things in this Light will be larger than they ought to be if the Rays went on by the Bodies in strait Lines, and that these Shadows have three parallel Fringes, Bands or Ranks of colour'd Light adjacent to them. But if the Hole be enlarged the Fringes grow broad and run into one another, so that they cannot be distinguish'd. These broad Shadows and

Fringes have been reckon'd by some to proceed from the ordinary refraction of the Air, but without due examination of the Matter. For the circumstances of the Phaenomenon, so far as I have observed them, are as follows.

Obs. 1. I made in a piece of Lead a small Hole with a Pin, whose breadth was the 42d part of an Inch. For 21 of those Pins laid together took up the breadth of half an Inch. Through this Hole I let into my darken'd Chamber a beam of the Sun's Light, and found that the Shadows of Hairs, Thred, Pins, Straws, and such like slender Substances placed in this beam of Light, were considerably broader than they ought to be, if the Rays of Light passed on by these Bodies in right Lines. And particularly a Hair of a Man's Head, whose breadth was but the 280th part of an Inch, being held in this Light, at the distance of about twelve Feet from the Hole, did cast a Shadow which at the distance of four Inches from the Hair was the sixtieth part of an Inch broad, that is, above four times broader than the Hair, and at the distance of two Feet from the Hair was about the eight and twentieth part [1/28] of an Inch broad, that is, ten times broader than the Hair, and at the distance of ten Feet was the eighth part of an Inch broad, that is 35 times broader.

The shadow-spreading, color-producing effects of diffraction are most evident when one uses small objects and light sources (here, the aperture is a rectangular slit about 0.024 inches, or 6 mm, wide). The disparity between the expected and the actual shadow becomes clearer if one reasons by similar triangles to discover what the breadth of the image would be if light traveled only in perfectly straight lines. At 4 inches from the hair the shadow would be about 1.2 times the breadth of the hair rather than 4 times; at 2 feet, about 2.25 times rather than 10 times; at 10 feet, about 7.4 times rather than 35 times. Thus the disparity gets proportionally larger at greater distances.

Nor is it material whether the Hair be encompassed with Air, or with any other pellucid Substance. For I wetted a polish'd Plate of Glass, and laid the Hair in the Water upon the Glass, and then laying another polish'd Plate of Glass upon it, so that the Water might fill up the space between the Glasses, I held them in the aforesaid beam of Light, so that the Light might pass through them perpendicularly, and the Shadow of the Hair was at the same distances as big as before. The Shadows of Scratches made in polish'd Plates of Glass were also much broader than they ought to be, and the Veins in polish'd Plates of Glass did also cast the like broad Shadows. And therefore the great breadth of these Shadows proceeds from some other cause than the Refraction of the Air.

As usual, the experiments Newton records are theoretically productive as well as demonstrative of important phenomena. By encompassing the hair with water without changing the outcome, he shows that the original phenomenon was not caused by refractive effects of the surrounding air.

Let the Circle X [in Fig. 33] represent the middle of the Hair; ADG, BEH, CFI, three Rays passing by one side of the Hair at several distances; KNQ, LOR, MPS, three other Rays passing by the other side of the Hair at the like distances; D, E, F, and N, O, P, the places where the Rays are bent in their passage by the Hair; G, H, I, and Q, R, S, the places where the Rays fall on a Paper GQ; IS the breadth of the Shadow of the Hair cast on the Paper, and TI, VS, two Rays passing to the Points I and S without bending when the Hair is taken away. And it's manifest that all the Light between these two Rays TI and VS is bent in passing by the Hair, and turned aside from the Shadow IS, because if any part of this Light were not bent it would fall on the Paper within the Shadow, and there illuminate the Paper, contrary to experience. And because when the Paper is at a great distance from the Hair, the Shadow is broad, and therefore the Rays TI and VS are at a great distance from one another, it follows that the Hair acts upon the Rays of Light at a good distance in their passing by it. But the Action is strongest on the Rays which pass by at least distances, and grows weaker and weaker accordingly as the Rays pass by at distances greater and greater, as is represented in the Scheme: For thence it comes to pass, that the Shadow of the Hair is much broader in proportion to the distance of the Paper from the Hair, when the Paper is nearer the Hair, than when it is at a great distance from it.

In contrast to what he had done in Books I and II, Newton presents the basic phenomenon and his explanation of what is happening at the same time. Forces are exerted on the rays of light as they pass the edge of the hair so that they are pushed away from the edge as they pass; this is a short-range force in that it has a very strong effect right next to the edge and rapidly decreases to zero as the distance increases.

Newton goes on to show that one gets the same results with other materials besides hair; he then proceeds to describe the phenomenon in detail. He begins using a knife blade in order to see whether the fineness or sharpness of the edge makes a difference and to record the results when the edge is long and even. He joins two blades in the manner of a pair of scissors to study what happens when light passes by two edges very close to one another. In this case he finds that one gets edge

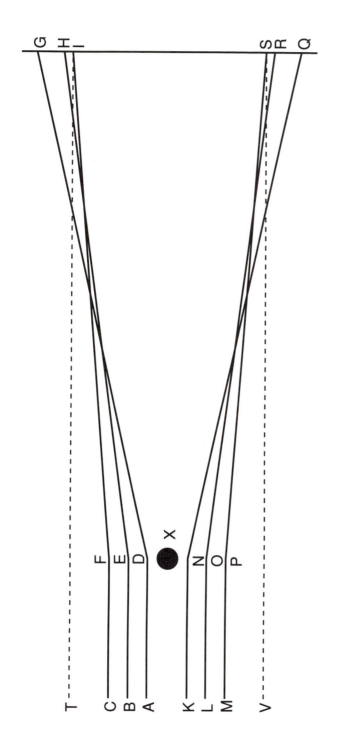

Fig. 33

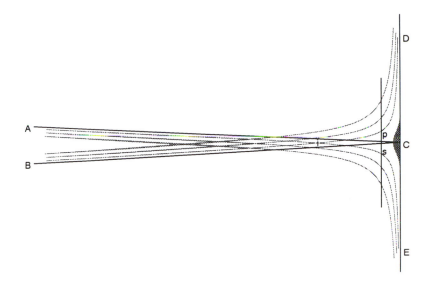

Fig. 34

shadows like those shown in Fig. 34 (ACB represents the perfect shadow the pivoted scissors blades would cast were the paths of the rays straight; C stands for the point of the perfect geometrical shadow where the two blades meet; to the left of C the gap between the blades increases to a maximum separation near AB). Notice that as one moves rightward from A and B the two sets of shadows are parallel to their respective edges, but when the blades are quite close to one another (near C) the "inflection" rapidly increases. What is happening, according to Newton, is that rays passing nearest the lower blade are deflected further and further upward, while the upper blade deflects rays passing near it to the same degree downward. In addition, as the gap becomes minuscule, a dark, bulging shadow appears near C; that is, near C the blades allow virtually no light to go straight ahead but deflect all rays either up or down. It is interesting that the text (in Observation 10) does not mention this bulge, in fact seems to overlook or deny it by saying only that a light "illuminates all the triangular space *ip*DE*s* comprehended by these dark lines, and the right line DE."[2]

This is very strange, for one might expect that as the rays pass close to both edges one force would tend to counterbalance the other, so that some light would travel straight onward, or at least nearly so. This should hold true for any normal, "Newtonian" force. Gravity, for example, is a force that drops off according to an inverse square law, so that twice as far away the effect is only one-fourth as great. If the inflective force were governed by, say, an inverse fourth-power law, doubling the distance would result in just one-sixteenth (1 over 2^4) the force (i.e., a $15/16$ reduction at twice the distance); an inverse eighth-power law would reduce the effect at twice the distance by $255/256$. By further increasing the exponent one can model forces that operate very powerfully at small distances but virtually not at all at larger ones. If the fields of two such forces operating in contrary directions overlap, however, they should tend to mitigate or decrease one another's effects. Newton does not call attention to this problem.

In the final observation of Book III (number 11), Newton uses homogeneal light to determine the differential effect on different colored lights. The result is that the least refrangible (red) light is the most inflectible, and the most refrangible (violet) the least inflectible. But the investigation of inflection is broken off rather than concluded, and the preliminary findings remain paradoxical. As has already been remarked, the inflecting force does not seem to be a classical "Newtonian" force. It is also odd that even homogeneal light produces several discrete fringes, for what could make rays of exactly the same kind show up in a certain interval, not at all in the immediately adjacent space, but then again in a further-removed interval where the next fringe appears? If *refraction* is caused by a force that affects violet-producing particles *more* strongly than red (violet is refracted more at the same angle of incidence), what kind of *inflective* force could affect violet *less* than red (since violet is less inflected than red)? If the rays differ in size or speed, it would seem that different accelerative forces ought to have the same relative effect on particles of the same kind.

At any rate, quite apart from the difficulties of conceiving what kind of force might be operating here, Newton believes that the experiments on which any theory must be based need to be performed and observed more exactly. He leaves these for others to do. More important to him at this point are certain questions that have been at least implicit in the preceding books of the *Opticks* and that are intended to orient future research. The Queries might be called *leading* questions in two senses: almost all of them suggest in their very formulation what the answer is, while also giving direction to further studies. They begin with a question that inflection has made especially pressing, the possibility that bodies exert a force (or *action*) on light.

. . . And since I have not finish'd this part of my Design, I shall conclude with proposing only some Queries, in order to a farther search to be made by others.

Query 1. Do not Bodies act upon Light at a distance, and by their action bend its Rays; and is not this action (*caeteris paribus* [other things being equal]) strongest at the least distance?

Qu. 2. Do not the Rays which differ in Refrangibility differ also in Flexibility;[3] and are they not by their different Inflexions separated from one another, so as after separation to make the Colours in the three Fringes above described? And after what manner are they inflected to make those Fringes?

Qu. 3. Are not the Rays of Light in passing by the edges and sides of Bodies, bent several times backwards and forwards, with a motion like that of an Eel? And do not the three Fringes of colour'd Light above-mention'd arise from three such bendings?

Qu. 4. Do not the Rays of Light which fall upon Bodies, and are reflected or refracted, begin to bend before they arrive at the Bodies; and are they not reflected, refracted, and inflected, by one and the same Principle, acting variously in various Circumstances?

These first four Queries direct the investigator to look for the causes of light's behavior in the forces that bodies exert on the rays. The study of thin-plate colors concluded that forces exist at or near body surfaces, and Grimaldi's diffraction effects suggested to Newton that edges exert forces as well. A ray traveling near to a refractive surface seems almost pulled into the new medium; one passing near an edge seems to be repelled. Thus it appears that rays of light in their encounters with matter are constantly turned from one path to another, in an "eel-like" motion. Boldest of all here is the fourth Query's suggestion that it might be one and the same principle (force?) that produces all such phenomena.

Queries 5 through 11 turn to the apparent interrelationship of light and bodies that gives rise to heat, a matter that was just barely touched upon in Proposition XI of Book II, Part III.

Qu. 5. Do not Bodies and Light act mutually upon one another; that is to say, Bodies upon Light in emitting, reflecting, refracting and inflecting it, and Light upon Bodies for heating them, and putting their parts into a vibrating motion wherein heat consists?

Qu. 6. Do not black Bodies conceive [take on] heat more easily from Light than those of other Colours do, by reason that the Light falling on them is not reflected outwards, but enters the Bodies, and is often reflected and refracted within them, until it be stifled and lost?

Qu. 7. Is not the strength and vigor of the action between Light and sulphureous Bodies observed above, one reason why sulphureous Bodies take fire more readily, and burn more vehemently than other Bodies do?

Qu. 8. Do not all fix'd Bodies, when heated beyond a certain degree, emit Light and shine; and is not this Emission performed by the vibrating motions of their parts? And do not all Bodies which abound with terrestrial parts, and especially with sulphureous ones, emit Light as often as those parts are sufficiently agitated; whether that agitation be made by Heat, or by Friction, or Percussion, or Putrefaction, or by any vital Motion, or any other Cause? . . .

Qu. 9. Is not Fire a Body heated so hot as to emit Light copiously? For what else is a red hot Iron than Fire? And what else is a burning Coal than red hot Wood?

Qu. 10. Is not Flame a Vapour, Fume or Exhalation heated red hot, that is, so hot as to shine? For Bodies do not flame without emitting a copious Fume, and this Fume burns in the Flame. . . .

Qu. 11. Do not great Bodies conserve their heat the longest, their parts heating one another, and may not great dense and fix'd Bodies, when heated beyond a certain degree, emit Light so copiously, as by the Emission and Re-action of its Light, and the Reflexions and Refractions of its Rays within its Pores to grow still hotter, till it comes to a certain period of heat, such as is that of the Sun? And are not the Sun and fix'd Stars great Earths vehemently hot, whose heat is conserved by the greatness of the Bodies, and the mutual Action and Re-action between them, and the Light which they emit, and whose parts are kept from fuming away, not only by their fixity, but also by the vast weight and density of the Atmospheres incumbent upon them; and very strongly compressing them, and condensing the Vapours and Exhalations which arise from them? . . .

Queries 12 through 16 concern a topic that the rest of the *Opticks* had little to say about but that is crucial to the perception of light and colors: the effects of light that give rise to vision. As always, Newton tries to unify new topics with his previous findings. Thus, for example, the ray of light strikes the retina and produces vibrations (reminiscent of the fits of Book II); the optic nerves, being more or less dense, are capable of transmitting vibratory motions a long distance; the differences in vibrations might give rise to the differences in colors and to their harmonic relations; and so forth. In raising these questions Newton may be implicitly competing with Descartes, who had presented not only a theory of light (with which Newton violently disagreed and the physical foundations of which, he argued in the *Principia,* led to gross contradictions of experience) but also a theory of what happens to the impulses of light within the body, the nerves, and the brain.

Qu. 12. Do not the Rays of Light in falling upon the bottom of the Eye excite Vibrations in the *Tunica Retina* [the retina]? Which Vibrations, being propagated along the solid Fibres of the optic Nerves into the Brain, cause the Sense of seeing. For because dense Bodies conserve their Heat a long time, and the densest Bodies conserve their Heat the longest, the Vibrations of their parts are of a lasting nature, and therefore may be propagated along solid Fibres of uniform dense Matter to a great distance, for conveying into the Brain the impressions made upon all the Organs of Sense. For that Motion which can continue long in one and the same part of a Body, can be propagated a long way from one part to another, supposing the Body homogeneal, so that the Motion may not be reflected, refracted, interrupted or disorder'd by any unevenness of the Body.

Qu. 13. Do not several sorts of Rays make Vibrations of several bignesses, which according to their bignesses excite Sensations of several Colours, much after the manner that the Vibrations of the Air, according to their several bignesses excite Sensations of several Sounds? And particularly do not the most refrangible Rays excite the shortest Vibrations for making a Sensation of deep violet, the least refrangible the largest for making a Sensation of deep red, and the several intermediate sorts of Rays, Vibrations of several intermediate bignesses to make Sensations of the several intermediate Colours?

Qu. 14. May not the harmony and discord of Colours arise from the proportions of the Vibrations propagated through the Fibres of the optic Nerves into the Brain, as the harmony and discord of Sounds arise from the proportions of the Vibrations of the Air? For some Colours, if they be view'd together, are agreeable to one another, as those of Gold and Indigo, and others disagree.

Qu. 15. Are not the Species of Objects seen with both Eyes united where the optick Nerves meet before they come into the Brain, the Fibres on the right side of both Nerves uniting there, and after union going thence into the Brain in the Nerve which is on the right side of the Head, and the Fibres on the left side of both Nerves uniting in the same place, and after union going into the Brain in the Nerve which is on the left side of the Head, and these two Nerves meeting in the Brain in such a manner that their Fibres make but one entire Species or Picture, half of which on the right side of the Sensorium comes from the right side of both Eyes through the right side of both optick Nerves to the place where the Nerves meet, and from thence on the right side of the Head into the

Brain, and the other half on the left side of the Sensorium comes in like manner from the left side of both Eyes. For the optick Nerves of such Animals as look the same way with both Eyes (as of Men, Dogs, Sheep, Oxen, &c.) meet before they come into the Brain, but the optick Nerves of such Animals as do not look the same way with both Eyes (as of Fishes, and of the Chameleon,) do not meet, if I am rightly inform'd.

Qu. 16. When a Man in the dark presses either corner of his Eye with his Finger, and turns his Eye away from his Finger, he will see a Circle of Colours like those in the Feather of a Peacock's Tail. If the Eye and the Finger remain quiet these Colours vanish in a second Minute of Time, but if the Finger be moved with a quavering Motion they appear again. Do not these Colours arise from such Motions excited in the bottom of the Eye by the Pressure and Motion of the Finger, as at other times are excited there by Light for causing Vision? And do not the Motions once excited continue about a Second of Time before they cease? And when a Man by a stroke [blow] upon his Eye sees a flash of Light, are not the like Motions excited in the *Retina* by the stroke? And when a Coal of Fire moved nimbly in the circumference of a Circle, makes the whole circumference appear like a Circle of Fire; is it not because the Motions excited in the bottom of the Eye by the Rays of Light are of a lasting nature, and continue till the Coal of Fire in going round returns to its former place? And considering the lastingness of the Motions excited in the bottom of the Eye by Light, are they not of a vibrating nature?

The color phenomena of Query 16 are caused not by light but by other causes producing a physiological response that leads to the perception of lights and colors. Blows to the eye, says Newton, might easily result in nerve vibrations that mimic the vibrations produced by light and lead to the perception of colors, and if the vibrations set off by light rays in the retina are at all enduring it would not be difficult to explain the existence of afterimages. Although Newton's physiological theories are fairly primitive and secondhand, they exploit what he has tried to establish earlier in the *Opticks*.

Queries 17–24 first appeared in the second English edition of the *Opticks* (1717), 13 years after the first edition. (Queries 25–31 had appeared earlier, in 1706, as numbers 17–23 of the first Latin edition.) With its discussion of vibratory motion Query 17 neatly picks up where Query 16 had left off; it attempts to draw an explicit connection between the vibrations in the eye and the theory of fits of easy transmission and reflection by way of an analogy.

Qu. 17. If a Stone be thrown into stagnating Water, the Waves excited thereby continue some time to arise in the place where the Stone fell into the Water, and are propagated from thence in concentrick Circles upon the Surface of the Water to great distances. And the Vibrations or Tremors excited in the Air by percussion, continue a little time to move from the place of percussion in concentrick Spheres to great distances. And in like manner, when a Ray of Light falls upon the Surface of any pellucid Body, and is there refracted or reflected, may not Waves of Vibrations, or Tremors, be thereby excited in the refracting or reflecting Medium at the point of Incidence, and continue to arise there, and to be propagated from thence as long as they continue to arise and be propagated, when they are excited in the bottom of the Eye by the Pressure or Motion of the Finger, or by the Light which comes from the Coal of Fire in the Experiments abovemention'd? and are not these Vibrations propagated from the point of Incidence to great distances? And do they not overtake the Rays of Light, and by overtaking them successively, do they not put them into the Fits of easy Reflexion and easy Transmission described above? For if the Rays endeavor to recede from the densest part of the Vibration, they may be alternately accelerated and retarded by the Vibrations overtaking them.

Query 18 sets out in a new direction: the investigation of the possibility of the existence of the ether, a very subtle matter that has rather extraordinary properties that enable it to serve as the medium responsible for the behavior of light (and in this sense the ether is already implicit in Query 17). The idea was not new to Newton, for he had presented a hypothesis about an ether in the "Hypothesis" letter concerning light and colors sent to the Royal Society in December 1675. The ether proposed in 1675 operated by collisions, pushes, and pulls; for example, there were supposed to be constant flows of ether toward the center of the earth that carried ordinary bodies along with them and thus gave rise to gravity. By the time of the *Principia*, however, Newton had settled upon a mathematical physics of particles giving rise to forces, at least as far as description was concerned; and although he provided mathematical accounts of how, for instance, a change in ether density at a surface might give rise to differential forces capable of producing refraction (owing to the repulsive forces exercised by the ether particles), he expressed reservations about committing himself to the actual existence of the ether. In the last years of his life Newton further developed this later notion of ether, both in his optics and in revisions of the *Principia*.

Newton begins Query 18 by describing phenomena that suggest the presence of some kind of matter even in what appears to be a vacuum.

Qu. 18. If in two large tall cylindrical Vessels of Glass inverted, two little Thermometers be suspended so as not to touch the Vessels, and the Air be drawn out of one of these Vessels, and these Vessels thus prepared be carried out of a cold place into a warm one; the Thermometer *in vacuo* will grow warm as much, and almost as soon as the Thermometer which is not *in vacuo*. And when the Vessels are carried back into the cold place, the Thermometer *in vacuo* will grow cold almost as soon as the other Thermometer. Is not the Heat of the warm Room convey'd through the *Vacuum* by the Vibrations of a much subtiler Medium than Air, which after the Air was drawn out remained in the *Vacuum?* And is not this Medium the same with that Medium by which Light is refracted and reflected, and by whose Vibrations Light communicates Heat to Bodies, and is put into Fits of easy Reflexion and easy Transmission? And do not the Vibrations of this Medium in hot Bodies contribute to the intenseness and duration of their Heat? And do not hot Bodies communicate their Heat to contiguous cold ones, by the Vibrations of this Medium propagated from them into the cold ones? And is not this Medium exceedingly more rare and subtile than the Air, and exceedingly more elastic and active? And doth it not readily pervade all Bodies? And is it not (by its elastick force) expanded through all the Heavens?

Heat, the earlier Queries suggested, is a vibration of the small parts of bodies. How, then, is heat transferred through a vacuum to the isolated thermometer if there is no matter to convey the heat? In response Newton presents the thesis that there is a substance consisting of particles much smaller, much rarer (i.e., more spread out), much more elastic (i.e., capable of restoring their shape or position when subjected to a collision or force; a billiard ball is more elastic than a marshmallow), and much more active than particles of ordinary matter, a substance that not only can communicate by its vibrations light's heating power to bodies but also serves as the medium of the chief phenomena of light: reflection, refraction, and fits of easy transmission and reflection.

In the next four Queries Newton develops the idea of the differential density of the ether. Dense bodies, that is, bodies that pack a great deal of ordinary matter into a unit volume, contain relatively little ether, whereas light bodies are richer in ether content, and so-called empty space is filled with ether. So equal volumes of empty space, when the empty spaces are contained *within* bodies—the "pores" in them—do *not* contain the same quantity of ether, at least whenever the bodies consist of "normal" matter having *different* densities. The more densely packed ordinary matter is, the more it *repels* ether particles from its pores. The empty spaces within two different bodies, their pores, will not contain equal quantities of ether, except when the solid parts of the two bodies are of equal density. (There is a kind of repulsive force between ether and regular matter.) This differential density in turn gives rise to "pressure gradients" in space: a difference in ether densities such as one finds at an interface between different bodies, like glass and air, gives rise to forces that account for refraction, gravity, and the like. In particular, since ether is rarer in dense bodies, light will have an easier time traveling through them; as Query 19 puts it, light always recedes from denser parts of the ether to rarer parts. This would help explain why when a ray of light enters a dense medium its path is refracted toward the perpendicular to the interface: the light has an easier time penetrating spaces where there is less ether.

Qu. 19. Doth not the Refraction of Light proceed from the different density of this Aethereal Medium in different places, the Light receding always from the denser parts of the Medium? And is not the density thereof greater in free and open Spaces void of Air and other grosser Bodies, than within the Pores of Water, Glass, Crystal, Gems, and other compact Bodies? For when Light passes through Glass or Crystal, and falling very obliquely upon the farther Surface thereof is totally reflected, the total Reflexion ought to proceed rather from the density and vigour of the Medium without and beyond the Glass, than from the rarity and weakness thereof.

Query 20 proposes that at interfaces the difference in ether density is not sharply defined (as though in crossing a line one instantaneously jumped from one density to another) but rather gradual—though perhaps across a distance small enough so that it seems sharply defined to the unaided human senses. This would imply that particles passing across such an interface would be subject to a gradual rather than an instantaneous acceleration over that whole distance, so that during this acceleration the particle might trace out a *curved* path. Query 21 then suggests that ether might even be the cause for gravitation.

Qu. 20. Doth not this Aethereal Medium in passing out of Water, Glass, Crystal, and other compact and dense Bodies into empty Spaces, grow denser and denser by degrees, and by that means refract the Rays of Light not in a point, but by bending them gradually in curve Lines? And doth not the gradual condensation of this Medium extend to some distance from the Bodies, and thereby cause the Inflexions of the Rays of Light, which pass by the edges of dense Bodies, at some distance from the Bodies?

Qu. 21. Is not this Medium much rarer within the dense Bodies of the Sun, Stars, Planets and Comets, than in the empty celestial Spaces between them? And in passing from them to great distances, doth it not grow denser and denser perpetually, and thereby cause the gravity of those great Bodies towards one another, and of their parts towards the Bodies; every Body endeavouring to go from the denser parts of the Medium towards the rarer? . . .

. . . .

Qu. 22. May not Planets and Comets, and all gross Bodies, perform their Motions more freely, and with less resistance in this Aethereal Medium than in any Fluid, which fills all Space adequately without leaving any Pores, and by consequence is much denser than Quick-silver or Gold? And may not its resistance be so small, as to be inconsiderable? For instance; If this *Aether* (for so I will call it) should be supposed 700,000 times more elastick than our Air, and above 700,000 times more rare; its resistance would be above 600,000,000 times less than that of Water. And so small a resistance would scarce make any sensible alteration in the Motions of the Planets in ten thousand Years. . . .

Query 22 removes one possibly major objection to the idea of the ether, especially to the notion that it is denser where space is otherwise empty. In such a case, all the planets would constantly, in their motion about the sun, be pushing their way through relatively dense concentrations of ether and thus should be slowed down. But, Newton responds, even a "dense" ether would exercise a vanishingly small resistance that would scarcely have a detectable effect on the tremendous mass of the planets.

Query 23 returns to the issue of vision, which had been taken up in Queries 12–17, but goes further by suggesting that the ether is the vibrating medium that conveys the motions of light through the body to the brain. Query 24 broaches a new topic: the ether may be the medium in living beings through which the will causes the body to move. Thus the ether not only fulfills physical and physiological functions but also plays a part in the psychological life of human beings, since it is conceived as the bodily medium through which desires and volitions are put into effect.[4]

Qu. 23. Is not Vision perform'd chiefly by the Vibrations of this Medium, excited in the bottom of the Eye by the Rays of Light, and propagated through the solid, pellucid and uniform Capillamenta [capillaries or small nerve fibers] of the optick Nerves into the place of Sensation? And is not Hearing perform'd by the Vibrations either of this or some other Medium, excited in the auditory Nerves by the Tremors of the Air, and propagated through the solid, pellucid and uniform Capillamenta of those Nerves into the place of Sensation? And so of the other Senses.

Qu. 24. Is not Animal Motion perform'd by the Vibrations of this Medium, excited in the Brain by the power of the Will, and propagated from thence through the solid, pellucid and uniform Capillamenta of the Nerves into the Muscles, for contracting and dilating them? I suppose that the Capillamenta of the Nerves are each of them solid and uniform, that the vibrating Motion of the Aethereal Medium may be propagated along them from one end to the other uniformly, and without interruption: For Obstructions in the Nerves create Palsies. And that they may be sufficiently uniform, I suppose them to be pellucid [perfectly transparent] when view'd singly, tho' the Reflexions in their cylindrical Surfaces may make the whole Nerve (composed of many Capillamenta) appear opake and white. For opacity arises from reflecting Surfaces, such as may disturb and interrupt the Motions of this Medium.

Queries 25 through 31 (which had been the concluding Queries 17–23 of the 1706 Latin edition) return to questions of light, beginning with a consideration of the phenomenon of double refraction in Query 25 but expanding in the course of Queries 30 and 31 to the general question of the interaction of light with matter and finally of the nature of matter and material creation.

Qu. 25. Are there not other original Properties of the Rays of Light, besides those already described? An instance of another original Property we have in the Refraction of Island-Crystal, described first by *Erasmus Bartholine,* and afterwards more exactly by *Hugenius,* in his Book *De la Lumiere.* . . .

In the *Traité de la Lumière* (composed around 1678 but not published until 1690) Christiaan Huygens had presented a rigorously mathematical account of the chief phenomena of light conceived as wave pulses in a highly elastic, etherial matter that entirely fills up space. In particular, he used this theory to explain the phenomenon of double refraction, which is exhibited by the crystal calcite, also known as Iceland spar. Erasmus

Bartholin (1625–1698) had in 1669 noted that in viewing an object through this material one sees a double image. He explained this as a result of the splitting of the incident ray into two refracted rays, the ordinary one, which follows the usual sine law of refraction, and the extraordinary ray, which goes off at a different angle. For example, when the angle of incidence on the crystal is 0°, that is, when the incident ray falls perpendicularly on the surface, the ordinary ray is not refracted, as would be expected for any transparent material when the incidence is perpendicular; the extraordinary ray, however, is refracted away from the perpendicular—which explains why it is called "extraordinary."

Huygens gave an ingenious and mathematically sophisticated account of this in terms of his wave theory, but it was based on several hypotheses about the mechanism of wave transmission that could not be verified.[5] Huygens also discovered that neither the extraordinary nor the ordinary ray is split if it enters a new Iceland crystal; but what is odder, if the ordinary ray enters the second crystal at certain orientations, it is refracted like an ordinary ray, but at other orientations like an extraordinary ray, and similarly for the extraordinary ray! Huygens had no explanation for this strange behavior. (It results from the transverse *polarization* of the light waves.)

It is clear why Newton had to take up the question of Iceland spar. First, once he decided to employ ether explanations, he had to differentiate his own theory of light from all other theories employing the concept. Huygens's ether filled space completely, the "particles" of it acted on one another by purely mechanical pressure and collision, and wave motions in it were equivalent to light; Newton's ether consisted of particles with some void between them, the particles acted more through force fields than through mechanical means, and they were the medium not of light, which itself was a kind of particle different from the ether, but of the interactions of light with ordinary matter. Second, Huygens's theory lent new legitimacy to modificationism, the theory that Newton believed he had once and forever laid to rest with his theory of refraction. Third, Huygens had employed a mathematics every bit as exacting as Newton's best work in the *Principia,* and so it might easily give the appearance of being a more rigorous theory of light than Newton's own.

Newton did not attack Huygens's theory directly, however. Instead, he presented the basic phenomena of Iceland crystal refraction using one and two crystals and then drew the inductive conclusion that there was a new property of light to reckon with.

. . . And therefore there is an original difference in the Rays of Light, by means of which some Rays are in this Experiment constantly refracted after the usual manner, and others constantly after the unusual manner: For if the difference be not original, but arises from new Modifications impress'd on the Rays at their first Refraction, it would be alter'd by new Modifications in the three following Refractions; whereas it suffers no alteration, but is constant, and has the same effect upon the Rays in all the Refractions. The unusual Refraction is therefore perform'd by an original property of the Rays. And it remains to be enquired, whether the Rays have not more original Properties than are yet discover'd.

The strategy is exactly the same one that Newton had followed in every encounter with modificationism: if you believe that a phenomenon comes about because an event modifies light, then every further occurrence of that event ought to cause the same modification to occur again; in this case, each ray ought to be divided again at each additional refraction in Iceland crystal. But using a second crystal shows that this does not always happen; therefore, the modificationist hypothesis is wrong. Newton's alternative is to suggest that a new property of light has been uncovered, and he speculates that the rays of light may have sides, or poles, that are revealed by the abnormal refraction in Iceland crystal. (The modern notion of polarization is understood in terms of the axis of orientation of the transverse waves.) He also cites (in Query 28) the authority of "the oldest and most celebrated Philosophers of *Greece* and *Phoenicia*" for rejecting the notion underlying Huygens's and Descartes's understanding of light, that space is entirely filled with a highly elastic matter (a plenum). This appeal to the ancients is rooted in Newton's conviction that the wise men of antiquity and the Hebrew prophets understood the fundamental truths of the sciences of nature, and so serves as a rhetorical appeal to authority tending to undermine confidence in the validity of any kind of modificationism.[6]

Qu. 26. Have not the Rays of Light several sides, endued with several original Properties? For if the Planes of perpendicular Refraction of the second Crystal be at right Angles with the Planes of perpendicular Refraction of the first Crystal, the Rays which are refracted after the usual manner in passing through the first Crystal, will be all of them refracted after the unusual manner in passing through the second Crystal; and the Rays which are refracted after the unusual manner in passing through the first Crystal, will be all of them refracted after the usual manner in passing through the second Crystal. And therefore there are not two sorts of Rays differing in their nature from one another, one of which is constantly and in all Positions refracted after the usual manner, and the other constantly and in all Positions after the unusual manner. The difference between the two sorts of Rays in the Experiment mention'd in the 25th Question, was only in the Positions of the Sides of the Rays to the Planes of perpendicular Refraction. . . .

Every Ray of Light has therefore two opposite Sides, originally endued with a Property on which the unusual Refraction depends, and the other two opposite Sides not endued with that Property. And it remains to be enquired, whether there are not more Properties of Light by which the Sides of the Rays differ, and are distinguish'd from one another.

. . . .

Qu. 27. Are not all Hypotheses erroneous which have hitherto been invented for explaining the Phaenomena of Light, by new Modifications of the Rays? For those Phaenomena depend not upon new Modifications, as has been supposed, but upon the original and unchangeable Properties of the Rays.

Qu. 28. Are not all Hypotheses erroneous, in which Light is supposed to consist in Pression or Motion, propagated through a fluid Medium? For in all these Hypotheses the Phaenomena of Light have been hitherto explain'd by supposing that they arise from new Modifications of the Rays; which is an erroneous Supposition.

If Light consisted only in Pression propagated without actual Motion, it would not be able to agitate and heat the Bodies which refract and reflect it. If it consisted in Motion propagated to all distances in an instant, it would require an infinite force every moment, in every shining Particle, to generate that Motion.[7] And if it consisted in Pression or Motion, propagated either in an instant or in time, it would bend into the Shadow. For Pression or Motion cannot be propagated in a Fluid in right Lines, beyond an Obstacle which stops part of the Motion, but will bend and spread every way into the quiescent Medium which lies beyond the Obstacle.[8] . . .

. . . .

And for rejecting such a Medium, we have the Authority of those the oldest and most celebrated Philosophers of *Greece* and *Phoenicia,* who made a *Vacuum,* and Atoms, and the Gravity of Atoms, the first Principles of their Philosophy; tacitly attributing Gravity to some other Cause than dense Matter. Later Philosophers banish the Consideration of such a Cause out of natural Philosophy, feigning [inventing] Hypotheses for explaining all things mechanically, and referring other Causes to Metaphysicks: Whereas the main Business of natural Philosophy is to argue from Phaenomena without feigning Hypotheses, and to deduce Causes from Effects, till we come to the very first Cause, which certainly is not mechanical; and not only to unfold the Mechanism of the World, but chiefly to resolve these and such like Questions. What is there in places almost empty of Matter, and whence is it that the Sun and Planets gravitate towards one another, without dense Matter between them? Whence is it that Nature doth nothing in vain; and whence arises all that Order and Beauty which we see in the World? To what end are Comets, and whence is it that Planets move all one and the same way in Orbs concentrick, while Comets move all manner of ways in Orbs very excentrick; and what hinders the fix'd Stars from falling upon one another? How came the Bodies of Animals to be contrived with so much Art, and for what ends were their several Parts? Was the Eye contrived without Skill in Opticks, and the Ear without Knowledge of Sounds? How do the Motions of the Body follow from the Will, and whence is the Instinct in Animals? Is not the Sensory of Animals that place to which the sensitive Substance is present, and into which the sensible Species of Things are carried through the Nerves and Brain, that there they may be perceived by their immediate presence to that Substance? And these things being rightly dispatch'd, does it not appear from Phaenomena that there is a Being incorporeal, living, intelligent, omnipresent, who in infinite Space, as it were in his Sensory, sees the things themselves intimately, and throughly [thoroughly] perceives them, and comprehends them wholly by their immediate presence to himself: Of which things the Images only carried through the Organs of Sense into our little Sensoriums, are there seen and beheld by that which in us perceives and thinks. And though every true Step made in this Philosophy brings us not immediately to the Knowledge of the first Cause, yet it brings us nearer to it, and on that account is to be highly valued.

Query 28 pulls out all the scientific and philosophical stops to put an end to the theory of light as wave motion. According to Newton, it is not just a hypothesis that can be proved wrong by experiment and experience, it is the wrong way of doing science, for it proceeds by making or feigning hypotheses rather than by acquiring genuine knowledge. Because it is founded on the notion that all hypotheses should be mechanical, it can never come to an understanding of nature as a whole, since, as Newton says, the first cause—that is, God—is not at all mechanical. Although collisions, forces, and the like may well explain many phenomena, mechanism by itself cannot account for its own validity or truth, nor by itself can it explain the dynamic order of the stars and planets, the perfection and purposeful functioning of animals and their bodies, and the nature of human willing, perceiving, and understanding. All these things point to a higher cause than physical laws can provide.

Query 29 resumes Newton's positive inquiry into what light is. It is a body of a particular kind, but as body it is like all other material things in nature and thus in constant interaction with them in accordance with the forces that govern all things.

Qu. 29. Are not the Rays of Light very small Bodies emitted from shining Substances? For such Bodies will pass through uniform Mediums in right Lines without bending into the Shadow, which is the Nature of the Rays of Light. They will also be capable of several Properties, and be able to conserve their Properties unchanged in passing through several Mediums, which is another Condition of the Rays of Light. Pellucid Substances act upon the Rays of Light at a distance in refracting, reflecting, and inflecting them, and the Rays mutually agitate the parts of those Substances at a distance for heating them; and this Action and Re-action at a distance very much resembles an attractive Force between Bodies. If Refraction be perform'd by Attraction of the Rays, the Sines of Incidence must be to the Sines of Refraction in a given Proportion, as we shew'd in our Principles of Philosophy: And this Rule is true by Experience. The Rays of Light in going out of Glass into a *Vacuum,* are bent towards the Glass; and if they fall too obliquely on the *Vacuum,* they are bent backwards into the Glass, and totally reflected; and this Reflexion cannot be ascribed to the Resistance of an absolute *Vacuum,* but must be caused by the Power of the Glass attracting the Rays at their going out of it into the *Vacuum,* and bringing them back. . . . Nothing more is requisite for producing all the variety of Colours, and degrees of Refrangibility, than that the Rays of Light be Bodies of different Sizes, the least of which may make violet the weakest and darkest of the Colours, and be more easily diverted by refracting Surfaces from the right Course; and the rest as they are bigger and bigger, may make the stronger and more lucid Colours, blue, green, yellow, and red, and be more and more difficultly diverted. Nothing more is requisite for putting the Rays of Light into Fits of easy Reflexion and easy Transmission, than that they be small Bodies which by their attractive Powers, or some other Force, stir up Vibrations in what they act upon, which Vibrations being swifter than the Rays, overtake them successively, and agitate them so as by turns to increase and decrease their Velocities, and thereby put them into those Fits. And lastly, the unusual Refraction of Island-Crystal looks very much as if it were perform'd by some kind of attractive virtue lodged in certain Sides both of the Rays, and of the Particles of the Crystal. . . . [T]his argues a Virtue or Disposition in those Sides of the Rays, which answers to, and sympathizes with that Virtue or Disposition of the Crystal, as the Poles of two Magnets answer to one another. And as Magnetism may be intended [intensified] and remitted, and is found only in the Magnet and in Iron: So this Virtue of refracting the perpendicular Rays is greater in Island-Crystal, less in Crystal of the Rock, and is not yet found in other Bodies. I do not say that this Virtue is magnetical: It seems to be of another kind. I only say, that whatever it be, it's difficult to conceive how the Rays of Light, unless they be Bodies, can have a permanent Virtue in two of their Sides which is not in their other Sides, and this without any regard to their Position to the Space or Medium through which they pass.

Following Newton's insistence in Query 29 that light is a body with fixed properties, the turn now of Query 30 to the transmutations of light into gross bodies and vice versa may look quite strange. How can the great enemy of modification theories of light suddenly become a "modificationist" where matter is concerned? There is no necessary contradiction, however. Modificationism had been devised to explain the changes in light as it undergoes encounters with matter, as in refraction, changes that nevertheless do not turn it into something other than light. On this level, Newton insisted that all experiments showed that light as light can be analyzed into parts and reaggregated according to its various properties but that those properties do not change (with the possible exception of its fits, but he did not assert categorically that fits were just properties of the rays of light). None of this rules out the possibility that light might in some of its interactions with matter be turned into something other than light, nor that other matter might be converted into light. Rather, as Queries 30 and 31 show in abundant detail, transformations, or transmutations, are constantly experienced everywhere in nature. Newton is saying that God has created the world so that there are certain kinds of matter present in it at different levels of complexity and that through certain determinate physical and chemical or alchemical processes things can be broken down—to a point and degree that perhaps only God can know—and then transformed into other things. This is a more flexible philosophy of nature than, say, the kind of mechanism that asserts that everything consists of nothing but tiny, indestructible atoms that collide into, and hook onto, one another in various ways. The task of the natural philosopher is not to (over)simplify the world by saying everything is ultimately of one kind, but rather to identify and investigate all the properties and processes that exist in the universe, so that its full scope and majesty might be appreciated. Although some of the terms of Queries 30 and 31 seem to us antiquated or even quaint, they still manage to convey the breadth of Newton's vision and the intensity of his search for a fundamental unity in nature.

. . . *Qu.* 30. Are not gross Bodies and Light convertible into one another, and may not Bodies receive much of their Activity from the Particles of Light which enter their Composition? For all fix'd Bodies being heated emit Light so long as they continue sufficiently hot, and Light mutually stops in Bodies as often as its Rays strike upon their Parts, as we shew'd above. I know no Body less apt to shine than Water; and yet Water by frequent Distillations changes into fix'd Earth, as Mr. *Boyle* has try'd; and then this Earth being enabled to endure a sufficient Heat, shines by Heat like other Bodies.

The changing of Bodies into Light, and Light into Bodies, is very conformable to the Course of Nature, which seems delighted with Transmutations. Water, which is a very fluid tasteless Salt, she changes by Heat into Vapour, which is a sort of Air, and by Cold into Ice, which is a hard, pellucid, brittle, fusible Stone; and this Stone returns into Water by Heat, and Vapour returns into Water by Cold. Earth by Heat becomes Fire, and by Cold returns into Earth. Dense Bodies by Fermentation rarify into several sorts of Air, and this Air by Fermentation, and sometimes without it, returns into dense Bodies. Mercury appears sometimes in the form of a fluid Metal, sometimes in the form of a hard brittle Metal, sometimes in the form of a corrosive pellucid Salt call'd Sublimate, sometimes in the form of a tasteless, pellucid, volatile white Earth, call'd *Mercurius Dulcis* [mercurous chloride]; or in that of a red opake volatile Earth, call'd Cinnaber [an ore of mercury]; or in that of a red or white Precipitate, or in that of a fluid Salt; and in Distillation it turns into a Vapour, and being agitated *in Vacuo,* it shines like Fire. And after all these Changes it returns again into its first form of Mercury. Eggs grow from insensible Magnitudes, and change into Animals; Tadpoles into Frogs; and Worms into Flies. All Birds, Beasts and Fishes, Insects, Trees, and other Vegetables, with their several Parts, grow out of Water and watry Tinctures and Salts, and by Putrefaction return again into watry Substances. And Water standing a few Days in the open Air, yields a Tincture, which (like that of Malt) by standing longer yields a Sediment and a Spirit, but before Putrefaction is fit Nourishment for Animals and Vegetables. And among such various and strange Transmutations, why may not Nature change Bodies into Light, and Light into Bodies?

Qu. 31. Have not the small Particles of Bodies certain Powers, Virtues, or Forces, by which they act at a distance, not only upon the Rays of Light for reflecting, refracting, and inflecting them, but also upon one another for producing a great Part of the Phaenomena of Nature? For it's well known, that Bodies act one upon another by the Attractions of Gravity, Magnetism, and Electricity; and these Instances shew the Tenor and Course of Nature, and make it not improbable but that there may be more attractive Powers than these. For Nature is very consonant and conformable to her self. How these Attractions may be perform'd, I do not here consider. What I call Attraction may be perform'd by impulse, or by some other means unknown to me. I use that Word here to signify only in general any Force by which Bodies tend towards one another, whatsoever be the Cause. For we must learn from the Phaenomena of Nature what Bodies attract one another, and what are the Laws and Properties of the Attraction, before we enquire the Cause by which the Attraction is perform'd. The Attractions of Gravity, Magnetism, and Electricity, reach to very sensible distances, and so have been observed by vulgar Eyes, and there may be others which reach to so small distances as hitherto escape Observation; and perhaps electrical Attraction may reach to such small distances, even without being excited by Friction.

· ·

The Parts of all homogeneal hard Bodies which fully touch one another, stick together very strongly. And for explaining how this may be, some have invented hooked Atoms, which is begging the Question; and others tell us that Bodies are glued together by rest, that is, by an occult Quality, or rather by nothing; and others, that they stick together by conspiring Motions, that is, by relative rest amongst themselves. I had rather infer from their Cohesion, that their Particles attract one another by some Force, which in immediate Contact is exceeding strong, at small distances performs the chymical Operations above-mention'd, and reaches not far from the Particles with any sensible Effect.

All Bodies seem to be composed of hard Particles: For otherwise Fluids would not congeal; as Water, Oils, Vinegar, and Spirit or Oil of Vitriol do by freezing; Mercury by Fumes of Lead; Spirit of Nitre and Mercury, by dissolving the Mercury and evaporating the Flegm [any watery, inodorous, tasteless substance obtained by distillation]; Spirit of Wine and Spirit of Urine, by deflegming and mixing them; and Spirit of Urine and Spirit of Salt, by subliming them together to make Sal-armoniac [ammonium chloride]. Even the Rays of Light seem to be hard Bodies; for otherwise they would not retain different Properties in their different Sides. And therefore Hardness may be reckon'd the Property of all uncompounded Matter. At least, this seems to be as evident as the universal Impenetrability of Matter. For all Bodies, so far as Experience reaches, are either hard, or may be harden'd; and we have no other Evidence of universal Impenetrability, besides a large Experience without an experimental Exception. Now if compound Bodies are so very hard as we find some of them to be, and yet are very porous, and consist of Parts which are only laid together; the simple Particles which are void of Pores, and were never yet divided, must be much harder. For such hard Particles being heaped up together, can scarce touch one another in more than a few Points, and therefore must be separable by much less Force than is requisite to break a solid Particle, whose Parts touch in all the Space between them, without any Pores or Interstices to weaken their Cohesion. And how such very hard Particles which are only laid together and touch only in a few Points, can stick together, and that so firmly as they do, without the assistance of something which causes them to be attracted or press'd towards one another, is very difficult to conceive.

. . . .

Now the smallest Particles of Matter may cohere by the strongest Attractions, and compose bigger Particles of weaker Virtue; and many of these may cohere and compose bigger Particles whose Virtue is still weaker, and so on for divers Successions, until the Progression end in the biggest Particles on which the Operations in Chymistry, and the Colours of natural Bodies depend, and which by cohering compose Bodies of a sensible Magnitude. If the Body is compact, and bends or yields inward to Pression without any sliding of its Parts, it is hard and elastick, returning to its Figure with a Force rising from the mutual Attraction of its Parts. If the Parts slide upon one another, the Body is malleable or soft. If they slip easily, and are of a fit Size to be agitated by Heat, and the Heat is big enough to keep them in Agitation, the Body is fluid; and if it be apt to stick to things, it is humid; and the Drops of every Fluid affect a round Figure by the mutual Attraction of their Parts, as the Globe of the Earth and Sea affects a round Figure by the mutual Attraction of its Parts by Gravity.

Since Metals dissolved in Acids attract but a small quantity of the Acid, their attractive Force can reach but to a small distance from them. And as in Algebra, where affirmative Quantities vanish and cease, there negative ones begin; so in Mechanicks, where Attraction ceases, there a repulsive Virtue ought to succeed. And that there is such a Virtue, seems to follow from the Reflexions and Inflexions of the Rays of Light. For the Rays are repelled by Bodies in both these Cases, without the immediate Contact of the reflecting or inflecting Body. It seems also to follow from the Emission of Light; the Ray so soon as it is shaken off from a shining Body by the vibrating Motion of the Parts of the Body, and gets beyond the reach of Attraction, being driven away with exceeding great Velocity.[9] For that Force which is sufficient to turn it back in Reflexion, may be sufficient to emit it. It seems also to follow from the Production of Air and Vapour. The Particles when they are shaken off from Bodies by Heat or Fermentation, so soon as they are beyond the reach of the Attraction of the Body, receding from it, and also from one another with great Strength, and keeping at a distance, so as sometimes to take up above a Million of Times more space than they did before in the form of a dense Body. Which vast Contraction and Expansion seems unintelligible, by feigning the Particles of Air to be springy and ramous [branching], or rolled up like

Hoops, or by any other means than a repulsive Power.[10] The Particles of Fluids which do not cohere too strongly, and are of such a Smallness as renders them most susceptible of those Agitations which keep Liquors in a Fluor [flux], are most easily separated and rarified into Vapour, and in the Language of the Chymists, they are volatile, rarifying with an easy Heat, and condensing with Cold. But those which are grosser, and so less susceptible of Agitation, or cohere by a stronger Attraction, are not separated without a stronger Heat, or perhaps not without Fermentation. And these last are the Bodies which Chymists call fix'd, and being rarified by Fermentation become true permanent Air; those Particles receding from one another with the greatest Force, and being most difficultly brought together, which upon Contact cohere most strongly. And because the Particles of permanent Air are grosser, and arise from denser Substances than those of Vapours, thence it is that true Air is more ponderous than Vapour, and that a moist Atmosphere is lighter than a dry one, quantity for quantity. From the same repelling Power it seems to be that Flies walk upon the Water without wetting their Feet; and that the Object-glasses of long Telescopes lie upon one another without touching; and that dry Powders are difficultly made to touch one another so as to stick together, unless by melting them, or wetting them with Water, which by exhaling may bring them together; and that two polish'd Marbles, which by immediate Contact stick together, are difficultly brought so close together as to stick.

And thus Nature will be very conformable to her self and very simple, performing all the great Motions of the heavenly Bodies by the Attraction of Gravity which intercedes those Bodies, and almost all the small ones of their Particles by some other attractive and repelling Powers which intercede the Particles. The *Vis inertiae* [power of inertia] is a passive Principle by which Bodies persist in their Motion or Rest, receive Motion in proportion to the Force impressing it, and resist as much as they are resisted. By this Principle alone there never could have been any Motion in the World. Some other Principle was necessary for putting Bodies into Motion; and now they are in Motion, some other Principle is necessary for conserving the Motion. For from the various Composition of two Motions, 'tis very certain that there is not always the same quantity of Motion in the World. For if two Globes joined by a slender Rod, revolve about their common Center of Gravity with an uniform Motion, while that Center moves on uniformly in a right Line drawn in the Plane of their circular Motion; the Sum of the Motions of the two Globes, as often as the Globes are in the right Line described by their common Center of Gravity, will be bigger than the Sum of their Motions, when they are in a Line perpendicular to that right Line [see Appendix D]. By this Instance it appears that Motion may be got or lost. But by reason of the Tenacity of Fluids, and Attrition of their Parts, and the Weakness of Elasticity in Solids, Motion is much more apt to be lost than got, and is always upon the Decay.[11] For Bodies which are either absolutely hard, or so soft as to be void of Elasticity, will not rebound from one another. Impenetrability makes them only stop. If two equal Bodies meet directly *in vacuo,* they will by the Laws of Motion stop where they meet, and lose all their Motion, and remain in rest, unless they be elastick, and receive new Motion from their Spring. If they have so much Elasticity as suffices to make them re-bound with a quarter, or half, or three quarters of the Force with which they come together, they will lose three quarters, or half, or a quarter of their Motion. And this may be try'd, by letting two equal Pendulums fall against one another from equal heights. If the Pendulums be of Lead or soft Clay, they will lose all or almost all their Motions; If of elastick Bodies they will lose all but what they recover from their Elasticity. If it be said, that they can lose no Motion but what they communicate to other Bodies, the consequence is, that *in vacuo* they can lose no Motion, but when they meet they must go on and penetrate one another's Dimensions. If three equal round Vessels be filled, the one with Water, the other with Oil, the third with molten Pitch, and the Liquors be stirred about alike to give them a vortical Motion; the Pitch by its Tenacity will lose its Motion quickly, the Oil being less tenacious will keep it longer, and the Water being less tenacious will keep it longest, but yet will lose it in a short time. Whence it is easy to understand, that if many contiguous Vortices of molten Pitch were each of them as large as those which some suppose to revolve about the Sun and fix'd Stars, yet these and all their Parts would, by their Tenacity and Stiffness, communicate their Motion to one another till they all rested among themselves.[12] Vortices of Oil or Water, or some fluider Matter, might continue longer in Motion; but unless the Matter were void of all Tenacity and Attrition of Parts, and Communication of Motion, (which is not to be supposed,) the Motion would constantly decay. Seeing therefore the variety of Motion which we find in the World is always decreasing, there is a necessity of conserving and recruiting it by active Principles, such as are the Cause of Gravity, by which Planets and Comets keep their Motions in their Orbs, and Bodies acquire great Motion in falling; and the cause of Fermentation, by which the Heart and Blood of Animals are kept in perpetual Motion and Heat; the inward Parts of the Earth are constantly warm'd, and in some places grow very hot; Bodies burn and shine, Mountains take Fire, the Caverns of the Earth are blown up, and the Sun continues violently hot and lucid, and warms all things by his Light. For we meet with very little Motion in the World, besides what is owing to these active Principles. And if it were not for these Principles, the Bodies of the Earth, Planets, Comets, Sun, and all things in them, would grow cold and freeze, and become inactive Masses; and all Putrefaction, Generation, Vegetation and Life would cease, and the Planets and Comets would not remain in their Orbs.

All these things being consider'd, it seems probable to me, that God in the Beginning form'd Matter in solid, massy, hard, impenetrable, moveable Particles, of such Sizes and Figures, and with such other Properties, and in such Proportion to Space, as most conduced to the End for which he form'd them; and that these primitive Particles being Solids, are incomparably harder than any porous Bodies compounded of them; even so very hard, as never to wear or break in pieces; no ordinary Power being able to divide what God himself made one in the first Creation. While the Particles continue entire, they may compose Bodies of one and the same Nature and Texture in all Ages: But should they wear away, or break in pieces, the Nature of Things depending on them, would be changed. Water and Earth, composed of old worn Particles and Fragments of Particles, would not be of the same Nature and Texture now, with Water and Earth composed of entire Particles in the beginning. And therefore, that Nature may be lasting, the Changes of corporeal Things are to be placed only in the various Separations and new Associations and Motions of these permanent Particles; compound Bodies being apt to break, not in the midst of solid Particles, but where those Particles are laid together, and only touch in a few Points.[13]

It seems to me farther, that these Particles have not only a *Vis inertiae* [force of inertia], accompanied with such passive Laws of Motion as naturally result from that Force, but also that they are moved by certain active Principles, such as is that of Gravity, and that which causes Fermentation, and the Cohesion of Bodies. These Principles I consider, not as occult [hidden] Qualities, supposed to result from the specifick Forms of Things, but as general Laws of Nature, by which the Things themselves are form'd; their Truth appearing to us by Phaenomena, though their Causes be not yet discover'd. For these are manifest Qualities, and their Causes only are occult. And the *Aristotelians* gave the name of occult Qualities, not to manifest Qualities, but to such Qualities only as they supposed to lie hid in Bodies, and to be the unknown Causes of manifest Effects: Such as would be the Causes of Gravity, and of magnetick and electrick Attractions, and of Fermentations, if we should suppose that these Forces or Actions arose from Qualities unknown to us, and uncapable of being discovered and made manifest. Such occult Qualities put a stop to the Improvement of natural Philosophy, and therefore of late Years have been rejected. To tell us that every Species of Things is endow'd with an occult specifick Quality by which it acts and produces manifest Effects, is to tell us nothing: But to derive two or three general Principles of Motion from Phaenomena, and afterwards to tell us how the Properties and Actions of all corporeal Things follow from those manifest Principles, would be a very great step in Philosophy, though the Causes of those Principles were not yet discover'd: And therefore I scruple not to propose the Principles of Motion above-mention'd, they being of very general Extent, and leave their Causes to be found out.[14]

Now by the help of these Principles, all material Things seem to have been composed of the hard and solid Particles above-mention'd, variously associated in the first Creation by the Counsel of an intelligent Agent. For it became him who created them to set them in order. And if he did so, it's unphilosophical to seek for any other Origin of the World, or to pretend that it might arise out of a Chaos by the mere Laws of Nature;[15] though being once form'd, it may continue by those Laws for many Ages. For while Comets move in very excentrick Orbs in all manner of Positions, blind Fate could never make all the Planets move one and the same way in Orbs concentrick, some inconsiderable Irregularities excepted, which may have risen from the mutual Actions of Comets and Planets upon one another, and which will be apt to increase, till this System wants a Reformation. Such a wonderful Uniformity in the Planetary System must be allowed the Effect of Choice. And so must the Uniformity in the Bodies of Animals, they having generally a right and a left side shaped alike, and on either side of their Bodies two Legs behind, and either two Arms, or two Legs, or two Wings before upon their Shoulders, and between their Shoulders a Neck running down into a Back-bone, and a Head upon it; and in the Head two Ears, two Eyes, a Nose, a Mouth, and a Tongue, alike situated. Also the first Contrivance of those very artifical Parts of Animals, the Eyes, Ears, Brain, Muscles, Heart, Lungs, Midriff, Glands, Larynx, Hands, Wings, swimming Bladders, natural Spectacles [eyes], and other Organs of Sense and Motion; and the Instinct of Brutes and Insects, can be the effect of nothing else than the Wisdom and Skill of a powerful ever-living Agent, who being in all Places, is more able by his Will to move the Bodies within his boundless uniform Sensorium, and thereby to form and reform the Parts of the Universe, than we are by our Will to move the Parts of our own Bodies. And yet we are not to consider the World as the Body of God, or the several Parts thereof, as the Parts of God. He is an uniform Being, void of Organs, Members or Parts, and they are his Creatures subordinate to him, and subservient to his Will; and he is no more the Soul of them, than the Soul of Man is the Soul of the Species of Things carried through the Organs of Sense into the place of its Sensation, where it perceives them by means of its immediate Presence, without the Intervention of any third thing. The Organs of Sense are not for enabling the Soul to perceive the Species of Things in its Sensorium, but only for conveying them thither; and God has no need of such Organs, he being every where present to the Things themselves. And since Space is divisible *in infinitum,* and Matter is not necessarily in all places, it may be also allow'd that God is able to create Particles of Matter of several Sizes and Figures, and in several Proportions to Space, and perhaps of different Densities and Forces, and thereby to vary the Laws of Nature, and make Worlds of several sorts in several Parts of the Universe. At least, I see nothing of Contradiction in all this.[16]

As in Mathematicks, so in Natural Philosophy, the Investigation of difficult Things by the Method of Analysis, ought ever to precede the Method of Composition. This Analysis consists in making Experiments and Observations, and in drawing general Conclusions from them by Induction, and admitting of no Objections against the Conclusions, but such as are taken from Experiments, or other certain Truths. For Hypotheses are not to be regarded in experimental Philosophy. And although the arguing from Experiments and Observations by Induction be no Demonstration of general Conclusions; yet it is the best way of arguing which the Nature of Things admits of, and may be looked upon as so much the stronger, by how much the Induction is more general. And if no Exception occur from Phaenomena, the Conclusion may be pronounced generally. But if at any time afterwards any Exception shall occur from Experiments, it may then begin to be pronounced with such Exceptions as occur. By this way of Analysis we may proceed from Compounds to Ingredients, and from Motions to the Forces producing them; and in general, from Effects to their Causes, and from particular Causes to more general ones, till the Argument end in the most general. This is the Method of Analysis: And the Synthesis consists in assuming the Causes discover'd, and establish'd as Principles, and by them explaining the Phaenomena proceeding from them, and proving the Explanations.[17]

In the two first Books of these Opticks, I proceeded by this Analysis to discover and prove the original Differences of the Rays of Light in respect of Refrangibility, Reflexibility, and Colour, and their alternate Fits of easy Reflexion and easy Transmission, and the Properties of Bodies, both opake and pellucid, on which their Reflexions and Colours depend. And these Discoveries being proved, may be assumed in the Method of Composition for explaining the Phaenomena arising from them: An Instance of which Method I gave in the End of the first Book.[18] In this third Book I have only begun the Analysis of what remains to be discover'd about Light and its Effects upon the Frame of Nature, hinting several things about it, and leaving the Hints to be examin'd and improv'd by the farther Experiments and Observations of such as are inquisitive. And if natural Philosophy in all its Parts, by pursuing this Method, shall at length be perfected, the Bounds of Moral Philosophy will be also enlarged. For so far as we can know by natural Philosophy what is the first Cause, what Power he has over us, and what Benefits we receive from him, so far our Duty towards him, as well as that towards one another, will appear to us by the Light of Nature. And no doubt, if the Worship of false Gods had not blinded the Heathen, their moral Philosophy would have gone farther than to the four Cardinal Virtues [prudence or wisdom, justice, courage or fortitude, and temperance]; and instead of teaching the Transmigration of Souls, and to worship the Sun and Moon, and dead Heroes, they would have taught us to worship our true Author and Benefactor, as their Ancestors did under the Government of *Noah* and his Sons before they corrupted themselves.

FINIS.

Part 4

After the *Opticks*

Chapter 10

The Aftermath of the Optics of Newton

Newton transformed the science of optics. It can be argued that before him there was no optics in the modern sense, only an extension of the ancient tradition of geometrical optics augmented by the discovery of the sine law of refraction and such phenomena as thin-film and diffractive effects. Newton fused the approach of ancient optics with a recognizably modern method of investigative experimentation. The result was novel: an optics concerned not just with the mathematical description of ray paths but also with the properties and diverse effects of the entities that traveled those paths. The rigorous exploitation of experimentation with the help of mathematics produced a *physical* optics, which studied the substance and dynamics of the thing called light, alongside geometrical optics, which studied the paths taken by light.

Simultaneously Newton gave birth to the science of colors. This may even have been his *prime* objective, at least in the period of his earliest experiments with lenses and prisms. In one sense he only established more clearly than Descartes that colors were not in things but in sensations, in what was experienced in consciousness (in the sensorium, as Newton called it), and so the quality *color* could no longer be considered a strictly physical subject. Bodies are not colored but simply have a disposition to reflect and absorb different kinds of rays differently, and the rays themselves are not colored but color-producing. This puts physics and psychology on opposite sides of a disciplinary divide.

In another sense, by using different degrees of refrangibility to argue that there was a strict and precise correlation between colors and rays, and by claiming in essence that rays were themselves bodies (and thus substances), he at least partially rehabilitated the notion that colors were objective. Recall that Descartes thought color was caused by the rotation of the particles of light-transmitting matter; that is, it was neither a substance (not a homogeneal ray) nor a fixed property of a substance but a variable modification of a substance (the rotational motion of a particle). For Newton there was an intrinsic, invariable characteristic of rays that produced color when the rays entered the eye and set in motion the physiological process of seeing. Although Newton had very little to say about the physiological, much less the psychological, aspects of vision, he provided a physical schema of rays impinging on the retina that gave future theorists a starting point from which to think about these problems: the notion that there is a strong correlation between the colors seen and the composition of the rays that enter the eye. Furthermore, by taking a few halting steps toward organizing and quantifying colors with his color circle and the hypothesis of the harmonic proportions in the spectrum, he spurred explorations of *color space*[1] and the further quantification of color. Thus even some of Newton's missteps proved to be fruitful in the long run.

On the other hand, Newton's theory of color was taken in the eighteenth century as an essentially comprehensive explanation of why and how colors appear as they do. This theory, indeed much of Newton's optics, was typically presented in a rigidly dogmatic fashion (and not infrequently with errors, for example the assertion that there were *just* seven colors). Although this did not entirely close off research, it discouraged work that seemed to contradict Newton, especially when it combined with the widespread adulation of the greatest philosopher of nature who had ever lived. Moreover, the framework that Newton had established, the theoretical way of seeing light as different kinds of rays separable by instruments like lenses, prisms, and thin films, became sufficiently conventionalized that it was difficult to do research into light or color according to any alternative scheme. This led, for example, to a relative neglect of the study of such phenomena as afterimages and simultaneous contrast, which are not explicable solely in terms of the effect of light rays but require in addition a grasp of the eye's functioning.

Newton's belief in a strict correlation between light and color survived in the wave theory of the nineteenth century, which interpreted white light as consisting of all the different wavelengths (or frequencies) indiscriminately mixed together. Wavelengths indeed provided a more exacting way of specifying a numerical correlation between radiation and color; but as things turned out, toward the end of the nineteenth century it was shown that the wave theory can be interpreted as implying that white light is undifferentiated, that it is the refracting material that induces a change giving rise to color—a new, mathematized version of the modificationist theories of color that Newton had attacked so vigorously throughout his life.[2] The study of the physiology and psychology of color perception has only slowly shaken off the legacy of understanding "real" color as that of homogeneal, monochromatic rays and discovered that there is a much looser connection between rays or wavelengths and perceived colors than Newton thought.[3]

A particularly famous "error" of Newton's was his contention that chromatic distortion could not be eliminated or even significantly reduced by using lenses compounded of various materials. In the 1730s, however, the German mathematician Leonhard Euler showed theoretically how a reduction might be effected, and in the 1750s the British optician John Dollond actually produced such a lens (apparently without influence from Euler).[4] One may view this development in different lights, of course. On the one hand, Newton's pronouncement did discourage opticians from pursuing achromatic lenses, and Dollond's invention was initially received with some incredulity because everyone "knew" that such a thing was not possible. On the other hand, Dollond used Newton's technique of chromatic ray analysis in order to calculate how aberration might be eliminated. Moreover, even if Newton had not written his discouraging words, it is by no means sure that an achromatic lens would have been produced much earlier (54 years elapsed between the first edition of the *Opticks* and the publication of Dollond's discovery).[5]

Newton's misstep concerning chromatic aberration has been magnified precisely by the very high standards that he set with the rest of his work. In the course of the eighteenth century Newton became a cultural symbol of genius and greatness. Accordingly, flaws in his work and his character tended to be overlooked or suppressed, so that when this error in one of his major works was irrefutably recognized it became necessary to invent an apology for him. It was said, for instance, that Newton had overlooked variations in spectral dispersion produced by different materials, but otherwise his theory was correct. The major problem with such apologies was that they erroneously minimized the degree to which Newton's theory rests on his conviction that refrangibility was an absolutely unchangeable characteristic of light and that refraction had nothing to do with the particular material of the refracting medium. As we have seen from the Queries, it was not the matter of glass or water that was responsible for reflection, refraction, and the like, but rather the ether in the pores of matter and the forces it produced. If all the rays have fixed properties leading to fixed effects, and if with one kind of prism red-producing rays R are affected by forces to degree a, violet-producing V to degree $a + b$, and intermediate green-producing G to intermediate degree $a + b/2$, then any other kind of prism that affects R to degree a and V to degree $a + b$ must affect rays G to degree $a + b/2$ in order to maintain the proportionality of force to effect—but we know that this is not strictly true.

Newton's attempt to coordinate the spectrum with the musical scale was no accident, then; it was based on the conviction that the degree of dispersion within the spectrum depends not on the material but on the unchangeable properties of rays. The only thing that changes is the amount of attraction or repulsion exerted on the rays, but the attractive and repulsive forces do not depend on the particular matter being employed (e.g., the kind of transparent material a prism is made of). Instead, it is the density of ether that counts. Thus, the optical forces ought to maintain a strict and unvarying proportion in their effects on the different kinds of rays, and dispersion must maintain a fixed proportion that depends not on the material of the prism but on the ether density differences at the interface. If with one prism the green-producing light is refracted so that it crowds the blue (that is, occupies a portion of the spectrum that would be blue were one using a prism of a different material), and with another it crowds the yellow instead, the principle that refrangibility is a fixed and entirely unchangeable property of *rays* and not also dependent on the different *materials* collapses. It is interesting and ironic that Newton's project of investigating the inner structure of matter by using light as a kind of probe did not flourish until the nineteenth century, when wave theorists began to study the precise modes of light's emission, transmission, absorption, and other behavior in different materials; that is, once scientists took seriously the idea that different materials do produce differential effects on light.

Situated between the choices of whether light was corpuscular or impulse-like, whether it consisted of particles or was the result of motions carried along in a transmitting medium, Newton opted for the former. Although he was often quite cautious in expressing his commitment to the corpuscularity of light, there is little doubt from his writings, and most especially from the *Opticks,* that he thought the wave or impulse interpretation was incompatible with the most elementary facts about light. Sometimes this amounted to a refusal to recognize the genuine merits of modificationism. For example, his writings after 1690 show no acknowledgment that Huygens's theory of secondary wavelets (published in 1690) provided a mechanism by which light might be conceived as going around corners like other wave forms yet producing a shadow because of the net cancellation of one wavelet by another.

When at the beginning of the nineteenth century the wave theory began displacing the corpuscular theory, the followers of the Newtonian system put up a stubborn resistance, but by the 1820s the wave was triumphant. Newton's rejection of all impulse theories would seem to be an even worse mistake than the one concerning achromatic lenses. Still, we need to recall that Newton believed he could accommodate wave-like phenomena in his corpuscularism, for example by means of the fits of easy transmission and reflection for thin-plate colors.[6]

The project of the *Opticks* was not just the study of light. Rather, the study of light opened the way to an intensive, fine-grained study of matter and its laws. If light was a physical body, a corpuscle, then it was susceptible to the kind of analysis of bodies in fields of force that had been so successful in the *Principia,* and it allowed speculation about what forces at the microscopic level were responsible for the phenomena of reflection, refraction, inflection, and the like. It also allowed speculation about the interaction of light with other matter, including the matter of the nerves in the eye, and it suggested an intimate connection between heat and light. In these respects Newton is the ancestor of modern physics, especially cosmology and high-energy physics, which join observations of the far reaches of space to experiments in the interaction of invisible particles, with the purpose of investigating entities, structures, forces, and events in the furthest recesses of matter.

Whether Newton is to be blamed or praised for this or that aspect of his science, whether his theories are like or unlike those we appeal to today, are questions whose answers will vary as our theories and our conceptions of the nature of science change. When we raise them we need to realize what we are asking and why. If we judge everything according to what we were taught in our high school or college physics courses, we will be thinking only in terms of current scientific opinion. Although in an important sense we cannot help doing this, and although it is certainly legitimate to ask how contemporary science views past work, when we do this we run the risk of implicitly assuming that the present is always right and the past always likely to be wrong. But if there is anything to be learned from the past of the sciences, it is that any present, even our own, does not possess the whole truth, and that in quite decisive respects some of the best "knowledge" of an age will turn out to be limited, or defective, or even wrong when it is viewed by a future age. It would be foolish to conclude that we should be skeptics about everything, but equally foolish to commit ourselves unreservedly to everything said in the name of contemporary science.

In order to strike a balance between these extremes we can be helped by looking to the history of the sciences and to the acknowledged greats, like Newton, to get a clearer notion of what has changed and what has remained the same. So, for example, we can come to see that in an important sense Newton, in his penchant for working alone and in great secrecy, is unlike the typical scientist of today, who must work as part of a research team and publish results as rapidly as they come; yet we can also see that by reading and criticizing the writings of his predecessors and disputing with contemporaries, Newton was working in a community nonetheless. We can come to see that Newton was wrong not just in details but in some of his central theses, yet that his experiments, his theories, and his prestige decisively set research along paths that are still viable today. No one, I believe, can read the Queries without recognizing that Newton was asking questions that foreshadow modern attempts to unify all the known forces of nature, and that his sense of the dignity of science and the majesty of nature more than measures up to that of modern scientific cosmology.

The *Opticks* is one of the most influential scientific books ever written—in fact, one might even say, simply one of the most influential books, period. Although in scientific terms the *Principia* probably did more to shape the ideal of comprehensively subjecting natural phenomena to rigorous mathematical analysis, the level of technical demands it made on readers assured that its direct influence would be restricted to a small audience of mathematician-physicists (it of course had a wider dissemination through popularizations). The *Opticks,* in contrast, presented more an experimental than a mathematical science, and where it did not offer proofs it made suggestions, proposed hypotheses, and offered speculations that go so far as to embrace the innermost constitution of matter and God's providential ordering of the universe. As such it served as a paradigm or model for experimental science, for how it ought to be organized and conducted, and it also made clear that one needed to understand the universe as a whole in order to understand its parts.

The *Opticks* inspired similar experimental methods in other sciences, both where new phenomena were being discovered and where familiar phenomena had not yet been reduced to systematic order (for example, in the developing sciences of electricity and chemistry). It showed that one could begin an orderly investigation of things by identifying and isolating the properties of very basic phenomena, a stage that did not necessarily require sophisticated mathematics or rigorous measurement. With ingenuity and experimental technique one could try to arrange phenomena in a way that would *compel* them to reveal their secrets. The object was to arrive not at various hypotheses about things but at the *theory* of them—that is, the true and therefore appropriate way of viewing the properties of things manifested by experience and experiment.

The speculations of the Queries suggested that carefully designed experimental investigations would ultimately lead to an understanding of the fundamental kinds of matter and their forces. By progressively taking greater care in measurement and making necessary calculations and other applications of mathematics to the phenomena, it would be possible to arrive at an exact understanding of nature and its laws. The deeper one's understanding of the phenomena became, the further one would penetrate into the mysteries of nature, and the closer one would come to being able to pose and answer questions about the most fundamental things of all. By means of this progress of inquiry, one would also be acquiring a fuller comprehension of the order, wisdom, and majesty of God's creation, so that not just the demands of natural knowledge would be met but also those of religion and morality. The intended relevance of the *Opticks* thereby ranged from the humblest facts about light to the deepest truths about the universe, humanity, and God.

Is there a continuing lesson here for us and for future generations? Some people think that science tends to encourage atheism, whereas others think that even today it leads to questions that are theological. Some think that science is perfectly justified in itself or by virtue of its practical applications, whereas others think that it must be part of a larger framework of meaning. Some would say that Newton's philosophical and theological concerns are simply tacked on to the scientific work, but others would claim that they are part of a seamless whole. Who is right and who is wrong in making claims like these is not easy to answer; certainly these claims are not matters to be decided by physics alone. The questions behind the claims are ones that occur to whoever takes seriously the achievements of science and the impulse to know. Although we cannot assert that Newton is the only guide in the search for meaning and truth, we can nevertheless look to him as someone who made his ambitious inquiries with vigor and persistence and with a keen sense of what was at stake. Whatever the outcome of debates about his experiments and theories, he and his writings will long continue to be relevant to the question of the meaning and truth of science.

Appendix A

Impulse and Wave Theories of Light

In the first decades of the nineteenth century Thomas Young (1773–1829) and Augustin Fresnel (1788–1827) gave the wave theory of light a form that enabled it to displace the corpuscular theory. Despite the qualifications by twentieth-century quantum theory it remains as one of the most powerful ways of theoretically unifying the phenomena of light.

The ancestors of this theory existing in Newton's day were far less powerful and not so mathematically sophisticated. In fact, they are better described as *impulse* theories than wave theories (the latter implies repeated, periodic impulses). The most widely known was that of René Descartes, who argued that space was entirely filled with matter (a plenum—which means no empty space, no vacuum). The kinds of matter were basically three, distinguished by size and function more than by their nature: the macroscopic kind that we can see, a microscopic kind that transmits light, and an even smaller microscopic kind that fills up all the gaps in and between the other two kinds.

Descartes understood light as originating in the pressure or impulse exerted by a luminous body on the second kind of matter. This pressure was transmitted outward instantaneously (that is, with infinite speed). The pressure's effects fanned outward but were strongest in a straight-line direction perpendicular to the surface of the luminous body. This conception first appeared in *Le Monde* (*The World,* ca. 1630–1633), which was not published until long after Descartes's death, but it was also present in two works published during his lifetime, in the essays appended to the *Discourse on Method* (1637), and in the *Principles of Philosophy* (1644). (In the *Meteorology,* one of the three essays appended to the *Discourse,* Descartes also advanced the theory of the spreading and coloration of refracted light in terms of the rotation of tiny spheres of light matter, mentioned in Chapter 2.)

Three of the major critics of Newton's February 1672 letter, Pardies, Hooke, and Huygens, were proponents of impulse theories. Pardies and Huygens had mechanistic theories built on Descartes's principles, whereas

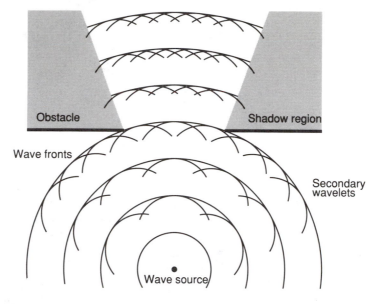

Fig. 35

Hooke, with a theoretical flexibility that marks his way of thinking, presented a variety of different mechanistic hypotheses of light as impulse phenomenon. In the *Micrographia* (1665) he suggested that light might be understood as a kind of impulse front that produced different colors depending on the front's orientation with respect to the direction of travel; refraction changes this orientation in white light so that colors appear. Hooke thought that similar hypotheses could explain the outcome of Newton's experiments and account for the apparent differential refrangibility according to color. He disputed Newton's contention that all the kinds of rays and colors had to be contained in white light even before refraction (since in his theory color results from a change in the wave front induced by the refraction).

In the early 1670s none of these men had a truly modern or mathematically sophisticated understanding of light as a wave, much less an understanding of how waves might be differentiated to account for differences in refraction and color. But by 1678 Huygens had taken a decisive step toward the modern theory by advancing a conception of how light can be understood as a wave front propagating itself outward in every direction from every point along the front by means of secondary wavelets (see Fig. 35). A star the size of a mathematical point is conceived as propagating a wave front simultaneously in all directions—the front would be like an expanding circle—and at every successive stage of propagation every point on the wave front circle acts as though it were a point source, radiating outward by means of the secondary wavelets.

A question raised by this scenario is why there are shadows, why a single source of light does not light up the whole universe from every direction, since each point of the successive wave fronts sends out impulses not just straight ahead but also to the sides (and perhaps even backwards)! Huygens's answer was that light was visible only where all the successive propagations reinforced one another, and mathematically this turns out to be precisely the expanding circle formed by the forwardmost part of each secondary wavelet (the so-called *envelope* of the wavelets). That is, all the portions of the impulse to the sides would cancel one another, for no net effect (Huygens showed this to be a consequence of the geometry of the wavelet situation), whereas the expanding circle surrounding the original source would be the envelope of visible light where all the wavelet fronts reinforce one another. Shadows are similarly a result of where the progressing wavefronts reinforce and cancel one another; the wavelets do go around corners, just like sound waves, but because of the geometry of wave propagation all the wavelets in the geometrical shadow cancel one another. Thus for the first time someone had shown that the notion of light as a wave or impulse effect was not inconsistent with one of the most basic phenomena of light, its propagation in straight lines.

According to Huygens's hypothesis light would travel more slowly in denser than in rarer materials (unlike Newton's theory, which required a greater velocity for light in denser materials), and the phenomenon of refraction was simply a mathematical consequence of the progressing wavefronts. By adding further hypotheses Huygens was also able to explain other phenomena, including the double refraction of calcite or Iceland spar (see the commentary on the *Opticks*, Book III, Part I, Query 25). Huygens's wavelet principle can be used to explain the diffraction of light (though Huygens himself did not do this). Huygens is also considered to be the discoverer of the polarization of light, although he did not himself provide a theory of its nature. The sum of his optical learning is contained in *De la Lumière* (*Treatise on Light*), published in 1690.

Huygens's theory is a wave *front* theory, because it does not at all depend on a continuous, rhythmic oscillation of the medium but rather only on the propagation of a single impulse, like one ripple spreading over the surface of a pond. It was not until the middle of the eighteenth century that Leonhard Euler (1707–1783) suggested that the differences in refrangibility and color so important to Newton could be accounted for by waves coming one after another in rapid succession, with different rates corresponding to different colors. Finally, in the early nineteenth century Young gave an explanation using the analogy of the propagation of sound waves (which are longitudinal waves consisting of a back-and-forth compression and rarefaction of air), and a decade later Fresnel provided a rigorous mathematization of light as consisting of transverse waves (which undulate up and down, perpendicular to the direction of propagation).

Some of the challenges that Huygens's new theory posed to the *Opticks* are explained in the main text. Huygens's theory had the major disadvantages of postulating an ether medium and not providing any explanation of what differences in light give rise to color, but on the other hand it was even more mathematically sophisticated and rigorous than Newton's. Like Newton's, it analyzed a visible phenomenon into a vast number of events taking place on a microscopic scale, but unlike Newton's it showed in meticulous detail precisely how the visible phenomenon was a strict mathematical consequence of the microevents.

Perhaps most important of all, it called into question the existence of the fundamental entity of Newton's theory, the ray of light. Consider the linear propagation of light: for Newton it is the simplest path for the ray to take (absent a deflecting force), but for Huygens it is the complex result of myriad secondary wavelets reinforcing and canceling one another; it just happens that the resulting light effects follow straight-line paths. For Newton shadows are areas to which the straight-traveling rays are denied access by boundaries of objects, but for Huygens wave disturbances travel around corners, only to be annulled by wavelet cancellation. Thus within the wave theory the term "ray" is just a manner of speaking, an imaginary line perpendicular to the real phenomenon, the advance of the wave. The "ray" is therefore a fiction that allows us to talk about an aspect of a complex event in oversimple terms. And whereas for Newton the ray's behavior is to be explained by the properties it possesses independent of any medium in which it travels, light for Huygens is a property of the medium, the medium's manner of transmitting pulses. If Huygens is right, then Newton's work needs to be reformulated right down to its very foundations, and even what Newton considers to be most certain, the doctrine of the differential refrangibility of rays, needs to be modified in view of the real mechanisms that produce the phenomena of light.

Appendix B

Estimating Air Thickness for the Colored Rings

In Book II, Part I, observation 5, Newton uses the square of the diameter of the colored rings that appear when a biconvex lens is pressed against the flat side of a planoconvex lens as an index for the "intervals of the rings," that is, the thickness of the plate or film of air between the curved and flat surfaces. This proportion is based on an approximation.

In Fig. 36, let the curve QODG represent the surface of a spherical lens with center of curvature P, and OAB the flat surface. From the figure (in which OA = CD and OB = EG) and the Pythagorean theorem (the square of the hypotenuse of a right triangle is equal to the sum of the squares of the sides, so that, in PCD, $PD^2 = CD^2 + PC^2$, and therefore $CD^2 = PD^2 - PC^2$; similarly $EG^2 = PG^2 - PE^2$) we get

$$OA^2/OB^2 = CD^2/EG^2 = (PD^2 - PC^2)/(PG^2 - PE^2)$$
$$= (PD + PC)(PD - PC)/(PG + PE)(PG - PE) \qquad (1)$$

Since they are all radii of the sphere, PG = PD = PO, so that, substituting for PG and PD in (1),

$$OA^2/OB^2 = (PO + PC)(PO - PC)/(PO + PE)(PO - PE) \qquad (2)$$

From the figure again, PO − PC = OC = AD, and PO − PE = OE = BG. Moreover, PO is approximately equal ($\approx$) to PC and PE, in particular when the curvature of the lens is gradual and A and B are taken very close to point O. Then PO + PC $\approx$ 2PO and PO + PE $\approx$ 2PO.

Substituting equals for equals in (2), we get

$$OA^2/OB^2 \approx (2PO \cdot AD)/(2PO \cdot BG) = AD/BG \qquad (3)$$

But this is what was wanted: When the curvature of the lens is gradual, that is, when its radius is large, and when the distances taken from O are not too

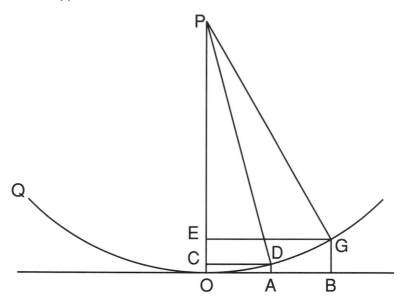

Fig. 36

great, the ratio of the squares of distances from O is about equal to the ratio of the corresponding gaps between the lens and the flat plate.

In observation 6 Newton uses another proportion: The diameter of the sphere of the lens curvature 2PO is to the semidiameter (radius) of the ring (OA) as this same semidiameter is to the air interval or gap (AD):

$$2PO/OA = OA/AD \text{ or } OA^2 = 2PO \cdot AD \tag{4}$$

By the Pythagorean theorem, $OA^2 = PA^2 - PO^2$. Also, $PA \approx PD + AD = PO + AD$. Consequently, once again by the Pythagorean theorem,

$$OA^2 \approx (PO + AD)^2 - PO^2 = PO^2 + 2PO \cdot AD + AD^2 - PO^2 = 2PO \cdot AD + AD^2 \tag{5}$$

AD, however, is a relatively small fraction of an inch, so that AD^2 will be a fraction of this fraction; in comparison to $2PO \cdot AD$ it is negligible. Therefore $OA^2 \approx 2PO \cdot AD$, which gives as approximately equal the relationship expressed in (4).

Appendix C

Rays in Force Fields

Proposition X of Book II, Part III, analyzes according to a force field model the refraction of a ray just grazing an interface. It is based on the idea of an accelerative force at the interface that changes the direction and speed of the ray and on the practice (already established in the generation before Newton) of dividing an actual motion into component motions along different axes.

A post-Newtonian way of explaining this has recourse to vectors. Vectors are arrows representing motion: length stands for speed, orientation indicates direction. Vector addition can then be defined in a way that mirrors the addition of motions: two vectors (motions) are added by putting the tail of one to the head of the other. For purposes of vector addition any vector of a given length and direction is equal to a parallel vector of the same length with the head pointed in the same direction. This means that you can slide the arrows about as long as you keep them oriented the same way. Moreover, as you can easily check by trying, the order in which vectors are added does not change the result.

More particularly, any motion represented by an arrow can be represented as the sum of two other arrows arranged *perpendicularly* to one another (see Fig. 37). Selecting the orientation of one of the axes in effect determines the length and direction of both arrows, since together they will form the sides of a right triangle with the original arrow as hypotenuse. There are innumerable ways of "analyzing" an arrow into perpendicular components, since axis orientation is a matter of convenience.

In Fig. 38, the grazing motion of the ray IC is almost entirely in the horizontal direction; thus if the total velocity (speed in a given direction) is represented by the segment IC, a division into vertical and horizontal components would yield a horizontal arrow virtually identical with IC and a vertical one that is nearly zero in length. Newton's central contention is that if the ray of light is a moving particle, then refraction is the result of an accelerative force applied to the particle at or near the refracting surface, a force that increases only the vertical component of motion but not the horizontal. (More than a generation earlier Descartes had noted this and illustrated it by conceiving refraction as similar to

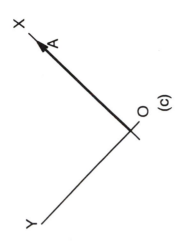

(c)

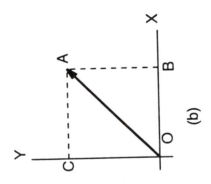

(b)

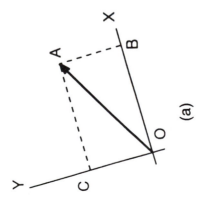

(a)

Fig. 37

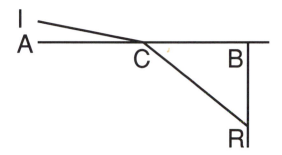

Fig. 38

the effect there would be if the ray were a tennis ball given a sharp vertical stroke by a racket precisely at the interface.) This force will leave the horizontal component of IC's motion unchanged, so that any difference in motion is attributable solely to the change in the vertical. Therefore BR (vector CR is the sum of BR and CB, where CB ≈ IC) is in effect a measure of the vertical acceleration produced by the refraction. In his analysis Newton points out that whatever the numerical law governing refraction might be, as long as there is some proportional relationship between distance and force (e.g., the inverse square law governing gravity), the total force applied to the particle in its passage through the entire field (conceived of as a series of parallel planes very close to one another) will be related to the added motion according to an equation of the form

$$k \sqrt{\text{(total force applied)}} = \text{(additional motion)}$$

where k is a constant number. Thus squaring the additional motion (that is, squaring BR) gives a direct proportional measure of the force. For mathematical details of this proof, the reader should look to the *Principia,* Book I, Proposition XCIV, Theorem XLVIII.

Having established this consequence of the force field interpretation of refraction, Newton then creates a chart showing that the experimentally determined values for the density of various refracting materials maintain roughly the same proportion as the square of the change in the vertical component of motion, BR. (That is, BR^2 divided by the material's density is roughly constant, whatever the material chosen.) The "roughness" of the correlation is expressed by the fact that the square of BR divided by density ranges from a relative value of 3,979 to 14,556, a difference of 3.66 times (if the correlation were perfect the values would be identical for all substances). For most glasses and crystals the difference is much less than this, however, and Newton suggests that the presence in ordinary bodies of unctuous or oily matter (sulfur was considered to be the preeminent oily matter) tends to increase refractive power.

Appendix D

The Changing Speed of Rotating Spheres

Newton makes a seemingly paradoxical assertion in Query 31. Suppose that two globes, joined by a slender rod, are moving as a whole in a straight line (i.e., their center of gravity traces a straight line) while they revolve around their common center of gravity. Newton says that whenever the two globes are oriented as in Fig. 39a, the total motion of the two-body system (i.e., the sum of the motions of the two bodies taken at that particular moment) will be greater than when the globes are oriented as in Fig. 39b. But given our understanding of the physics of this motion, an understanding that in fact derives from Newton himself, this is not so: in both cases the motion of the system is the same.

What Newton has in mind here is the sum of the speeds. In Fig. 39a, the speed of sphere 1 is $V - r$ to the right, and that of sphere 2 is $V + r$, also to the right. If we consider the motion of the whole to be the sum of the motions of the parts, then (neglecting the connecting rod) the total motion is $(V - r) + (V + r) = 2V$. In Fig. 39b, sphere 1 is moving V to the right and r downward, which means that the motion is really along a diagonal down and to the right, and its speed is determined by the Pythagorean theorem as $\sqrt{(V^2 + r^2)}$. Similarly, the speed of sphere 2 will be $\sqrt{(V^2 + r^2)}$ to the upper right. Summing the speeds without regard to direction, we get $2\sqrt{(V^2 + r^2)}$ which will always be greater than $2V$, the sum of the speeds in Fig. 39a.

If one reckons with velocity as a directed motion, however, in both cases the r components of the spheres will cancel (since they are 180° opposite one another), and the sum will be $2V$ to the right. So although the sum of the directed motion of the globes is always the same, the sum of the speeds without respect to the direction of motion is constantly changing.

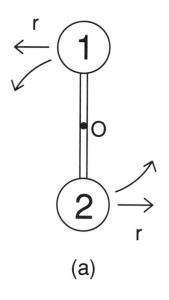

Motion of
center of gravity
of system (V)

$\longrightarrow$

(a)

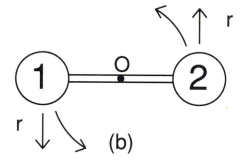

Motion of
center of gravity
of system (V)

$\longrightarrow$

(b)

Fig. 39

Notes

Full bibliographic data for items referenced in these notes by author's last name and short title can be found in the "Select Bibliography," to be found below.

Chapter 2

1. If I merely mention this work here it is not because of any lack of importance or intrinsic interest but rather because we have assembled elements sufficient for appreciating the context of Newton's work. For an examination of the optics of light and vision before 1600 see Lindberg, *Theories of Vision,* and Smith, "Getting the Big Picture."

2. The trigonometric functions give a way of relating the angles of a triangle to the relative proportions of the sides. The values of the sine, cosine, and tangent (and thus also their inverses, the cosecant, secant, and cotangent, respectively) can be determined as follows. Take a point anywhere along one of the sides of the angle Θ and drop a perpendicular to the other side to form a triangle (see Fig. 1). Let o indicate the side of this triangle opposite Θ. Let h indicate the hypotenuse of the triangle (the side opposite the right angle), and a the other side, adjacent to Θ. Then $\sin\Theta = o/h$, $\cos\Theta = a/h$, and $\tan\Theta = o/a$ (and also $\sin\Theta/\cos\Theta$). Since the hypotenuse is always longer than either of the two sides, the values of the sine and the cosine never exceed 1. The values do not depend on the actual size of the triangle drawn, since it is a question of the proportions between the sides.

3. Density is simply the mass (or, less precisely, the weight) of a material per unit of volume. For example, gold, at 19.3 g/cm^3 is about 80 times denser than cork, at 0.24 g/cm^3.

4. Of course they do not exhaust it. By the middle of the seventeenth century optical investigators were beginning to attend to phenomena such as diffraction and interference effects, which could not be reduced to reflection and refraction. But in a first attempt at understanding the behavior of light it

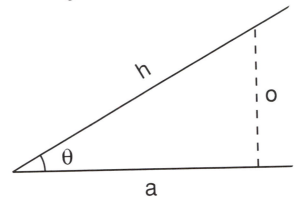

Fig. 1

is sufficient to appeal to the laws that govern how light "bounces off" some things (reflection, which comes into play both with smooth surfaces, like mirrors, and in a more complicated way with rough ones, that is, the majority of surfaces) and how it passes through others (refraction).

5. Some of the medieval approaches had required considering the action of different rays and determining which was predominant—the others could then be ignored.

6. The notebook, known as "Questiones quaedam philosophicae" (certain philosophical questions) and catalogued by the University of Cambridge Library as manuscript ULC Add. 3996, has been published, with commentary, in McGuire and Tamny, *Certain Philosophical Questions.* This volume also includes the somewhat later optical notes of ULC Add. 3975.

7. McGuire and Tamny, *Certain Philosophical Questions,* 241.

8. Newton assisted Barrow in preparing the latter's lectures for publication.

9. Newton was not the first to conceive of the idea of a telescope using mirrors rather than lenses—James Gregory (1638–1675) anticipated him by a few years—but he was the first to produce mirrors good enough for telescopic observation.

Chapter 3

1. See the letter of 18 January 1672, Newton, *Correspondence,* 1:82–83.

2. An actual spectrum is more likely to be pear- or teardrop-shaped, widest in the yellow and narrowing very considerably at the violet end. Newton's description accords more perfectly with his theory than with observation, and this might help explain why some contemporary critics had difficulties confirming Newton's results. For a discussion of some of the ways in which Newton corrected and idealized the experimental phenomena, see Lohne, "*Experimentum Crucis,*" and Laymon, "Newton's *Experimentum Crucis.*"

3. Imagine that a knife blade cuts straight through a perfect sphere. Each part now has a flat or plane side (the plane of the knife cut) and a convex curved side; that is, each is a planoconvex "lens" (in quotes because at this point it is a mathematical, not a real, object). The curve of this lens also can serve as the shape for a *planoconcave* lens if one imagines slicing a block of glass with a perfectly spherical bubble in the center. One can put together such curves in different ways: for instance, a biconvex lens (like most magnifying glasses) bulges outward on both sides, and a concavoconvex lens (often used to correct farsightedness) is hollow on one side and bulging on the other.

4. Note that this line of reasoning presupposes (rather than proves) that the total effect is simply a sum of the parts.

5. Note that the reasoning in this paragraph is based on a hypothetical model: the beam of light is conceived as a bundle of rays, all the rays are assumed to be refracted independently of one another according to the law of sines, and the outcome is described according to what the mathematics of the sine relationship implies. None of this determines what will actually happen when one tries the experiment; it only sets up an expectation deduced from the law and the initial conditions that are assumed. The assumption that the beam consists of rays is itself a hypothesis in need of confirmation. The tradition of representing the paths of light by mathematical lines does not in itself guarantee that real analogues of these lines exist—for example, in the wave theory rays are fictions, useful for descriptive purposes but not corresponding to the real causes (see Appendix A).

6. This is a point that has frequently been made in the secondary literature on this letter, for instance Sabra, *Theories of Light,* 235–236, and Guerlac, "Can We Date Newton's Early Optical Experiments?"

7. Newton coined this word, which has been the standard name ever since.

8. Newton's quick refutations do not entirely undermine the hypotheses, however. For example, the path length hypothesis could be formulated in terms of the fact that the side of the light beam nearer the prism's base will always traverse a slightly longer path in the glass, whereas a proponent of the hypothesis that the light must be delimited or bounded in some way could point out that whether the prism is in front of it or behind it, the hole still delimits the light; moreover, the faces of the prism delimit the light, and even the fact that the sun is a small, bright circle against the relatively dark background of the sky implies a delimiting of the light. Moreover, the proportions of the length to width in the spectrum *will* vary as the size of the hole is varied, so Newton perhaps should have explained more clearly what would have made these kinds of circumstances *material,* as he termed it.

9. The hypothesis appears to be one formed by Newton from hints provided by Descartes (in his account of how refraction might make a comet appear to have a tail) and Hooke. See Newton, *Correspondence,* 1:178.

10. Newton cannot have allowed the image to become very elongated before refracting it with the second prism, because with a prism of modest size not all the light of a fully elongated (and fully colored) spectrum would be intercepted, and also because excessive elongation would not be undone by an identical, inverted prism.

11. Measured from its center, a circle contains 360 degrees (360°); each degree can be divided into 60 minutes (60′); each minute, in turn, can be divided into 60 seconds (60″).

12. The angle subtended by an object is determined by the portion of the horizon that the object appears to occupy. A small object nearby subtends the same angle as a much larger one farther away (you can block out the sun with a coin held at arm's length). In the present case, the sun's breadth occupies 31′, or about 1/2°, of the horizon (about 1/720 of the whole celestial circle, which contains 360°), and so the corresponding image formed in the dark room by the unrefracted light of the sun on a screen 22 feet distant from the shutter aperture would occupy 1/720, or again 1/2°, of a circle of radius 22 feet (with the aperture as the circle's center).

13. The historical glass-thickness hypothesis was an explanation more of the variation of colors in the spectrum than of the spectrum's length. The shadow/bounding hypothesis tried to explain both elongation and color by means of a spreading edge disturbance. The scattering hypothesis was better suited to accounting for the elongation than for the color. Despite Newton's later protestation that in the first part of the letter he was concerned only with the geometry of elongation and not with color, it is not at all clear that these two aspects can be sharply separated from one another. For example, when Newton says that the image remained the same despite variations in the circumstances of the experiments, he does not distinguish between the two, but with certain variations there can be quite marked differences in the one but not the other (e.g., if the prism is placed outside the shutter one will get just a partial spectrum of colors if the prism is not almost touching the hole). For a fuller discussion see Sepper, *Goethe contra Newton,* 114–115.

14. Descartes, *Discourse on Method,* 334–339. Descartes did not conceive of the globules as moving through space, however, only as spinning in their more or less fixed positions.

15. Newton, *Correspondence,* 1:96–97.

16. "Crucial experiment" is a term that had been used by the natural philosopher Robert Hooke in the *Micrographia,* in a section dedicated to the investigation of light. The term has its roots in the *instantiae crucis,* instances of the crossing, described by Francis Bacon in his methodology of

experimental knowledge, the *New Organon*. The ancestry of the term is discussed further in Lohne, "*Experimentum Crucis*," 179.

17. See aphorism 36 of book 2 of the *New Organon*.

18. This term as used by Newton corresponds both to "kind" or "type" and also to "characteristic appearance"; thus red is a species of color insofar as it is a kind, but it is also a species (and the kind of color it is) because of its distinctive appearance—precisely as red!

19. This paragraph was not numbered in the original, but the preceding and following were numbered "3" and "5," respectively.

20. Using colored lights, one would usually call blue, green, and red the primaries; with dyes or ink, one would usually choose cyan, magenta, and yellow. The members of each set can be added to one another in varying amounts to produce all, or nearly all, other colors. Newton's use of "primary," by contrast, simply tells us that the color is produced by a single kind of ray.

21. The harder one presses the theory the more complications one finds. For example, even with modern equipment one would be hard pressed to isolate more than a hundred or so distinct hues in a given solar spectrum, much less "an indefinite variety" of intermediate gradations; see Raman, *The Physiology of Vision,* Chapter 8. About the "very precise and strict" analogy between color and refrangibility, Newton is somewhat less insistent in the *Opticks*. In a given spectrum one sees fairly broad bands of the same hue rather than a constant variation (in which equal changes in position would yield equal changes in hue). The addition to one another of different homogeneal colors also corresponds less well with Newton's claims in the green-to-violet portion of the spectrum than in the red-to-green; see Lohne, "*Experimentum Crucis*," 173. Nor had Newton worked out in 1672 the color circle that in the *Opticks* would be used to predict the colors that result from mixing rays.

22. The rainbow will be discussed below in the commentary on *Opticks, Book I, Part II.*

23. The wood of a small Mexican shrub. If slices of it are soaked in water, the resulting liquid will appear golden when light shines through it but sky blue in reflected light. This blue color is now understood to be due to fluorescence (a phenomenon that Newton's theory was not equipped to explain).

24. One point at issue is whether the experiment can be considered a success if where lens MN intercepts the refracted light there is not yet a full spectrum visible. With the size lenses Newton used it is more likely that the light was still chiefly white with narrow fringes of color. See Sepper, *Goethe contra Newton,* 154–156.

25. Newton, *Papers and Letters,* 59.

Chapter 4

1. See the letter of 13 April 1672 responding to Ignace Pardies; Newton, *Correspondence,* 1:142.

2. On the "etiquette" of claims to certainty, see Zev Bechler, "The Grammar of Scientific Dissent."

3. Most of this correspondence was published in the *Philosophical Transactions;* it can be found in Newton, *Papers and Letters,* and both the published and the unpublished letters are in Volumes 1 and 2 of Newton, *Correspondence.*

4. This does not imply that the thing is absolutely and without qualification fundamental but only with respect to the process of analysis. Chemical processes break matter down into atoms, but atoms are fundamental only with respect to chemical processes: nuclear processes of course can break atoms down further. In the phenomenon of fluorescence photons are absorbed and then reemitted at a different wavelength—thus unlike refraction and reflection fluorescence does change the ray.

5. In ordinary speech we tend to talk about the wavelength of monochromatic light, but since the length of the wave changes with the medium, it would probably be better to talk about frequency instead, which is the same in all media.

6. See Appendix A on impulse and wave theories available to Newton. Newton himself argued for the existence of an ether to explain gravitation, but unlike the ether of the wave theory his version did not completely fill up space and was not the medium of light transmission.

7. With the exception of a section of the *Principia* that explained how an ether might cause light corpuscles to change their paths in reflection and refraction; see Book I, Section XIV, Propositions XCIV–XCVIII (Newton, *Principia,* 1:226–233).

Chapter 5

1. On the history of Newton's optical writings and research, see Westfall, *Never at Rest,* esp. 238–280, 520–524, and 638–648.

2. In the 1670s the Danish astronomer Olaus Roemer had noticed that the eclipses of Jupiter's moons occur later than expected when the earth is farther away in its orbit from Jupiter and earlier when it is closer, and that this provides a way to calculate the speed of light, which had been thought by many to be infinite.

3. See McGuire and Tamny, *Certain Philosophical Questions,* 430–435.

4. The modificationists were not completely helpless before the phenomenon of repeated cross-refractions; see Sepper, *Goethe contra Newton,* 135–139.

5. I omit experiments 6–10 under Proposition II and the assertion in Proposition III that the sun's light consists of rays differing in *reflexibility,* a property of light that is more fully discussed in Book II, Part III. Experiment 6 is in essence like the experimentum crucis, though Newton assigns it no special proof value here in the *Opticks;* it instead provides instruction in how to obtain relatively homogeneal light in a practicable way. Experiment 7 uses two prisms to produce two spectra, the light from which is cast on white paper and thread observed through a third prism (with results similar to those of experiment 1), then superimposes the two spectra and reseparates them by again observing through another prism. Experiment 8 produces an analogue to experiment 2: prismatic colors are cast onto a printed page, and one ascertains that the distance at which a magnifying lens gives a readable image varies with color. Experiments 9 and 10, which support Proposition III, take advantage of the fact that at angles of incidence very close to the critical angle a prism will totally reflect violet and blue but transmit the other colors.

6. In Fig. 12, to the right of the aperture the area between lines *cb* and *ad* is illuminated by light from all points of the sun. Beyond *ab* and *cd* there is complete shadow, or *umbra*. And in the areas between *ab* and *cb* on one side and *ad* and *cd* on the other there is partially illuminated shadow, or *penumbra*.

7. This says that if the spectrum has length L and breadth B, then the degree of mixture of the light is given by the formula $B/(L - B)$. This leads to the anomaly that a spectrum twice as long as it is broad would be considered to be mixed to the degree $1/(2 - 1) = 1$, that is, the same as an unrefracted beam of sunlight. It is more plausible to use B/L (in which case the abovementioned spectrum would be only half as mixed as the ordinary beam).

8. The story is actually more complicated than this, as the subsequent history of optics shows. As will become clearer from Part II of Book I, Newton believed that the separation of the colors in the spectrum was a simple (linear) function of the degree to which the light was refracted out of its original path. One of the implications of this is that prisms made of different materials that produced the same degree of refraction should produce identical spectra. Consider two prisms, one made of glass and the other of plastic, each with a refracting angle of 40°. Since glass has a higher refractive index than plastic, the glass will refract the light more than the plastic (that is, the degree to which the light will be turned out of its original path is greater with glass than with plastic). But if you increase the refracting angle of the plastic sufficiently, say to 50°, the glass and the new plastic prism will refract the light to equal degrees. According to Newton's conception, the spectra of the 40° glass and the 50° plastic will be identical: if you

put them side by side the colors will match exactly. *But this is not in general true.* The colors will not match exactly because different transparent materials do not *disperse* all the colors in identical fashion. As a result, it is possible to assemble compound spherical lenses that reduce (and for certain frequencies of radiation eliminate) chromatic aberration, and thus much better refracting telescopes can be made than Newton thought possible. It was not until a generation after Newton's death that the British optician John Dollond found the practical solution to this problem (the German physicist Leonhard Euler had raised the theoretical possibility a decade earlier).

9. The editor of Newton's mathematical papers, Derek Whiteside, offers RR-RI as a correction for the RK-RI found in all editions, since the latter results in an impossibly simple equation. I, R, and K are numbers representing by their proportions to one another the relative refractive indices (I/R gives the index from glass into air, K/R from water into air, and, as a simple consequence of these two facts, I/K the index from glass into water). Newton is saying that if the outer, convex part of the glass lenses has been ground to a curvature corresponding to the surface of a sphere having diameter *d,* the inner, concave part of the lenses needs to be ground to a curvature corresponding to the surface of a sphere having diameter *e,* where

$$e = d \times \sqrt[3]{K^2 - KI} \Big/ \sqrt[3]{R^2 - RI}$$

A lens designed to these specifications would, according to Newton, eliminate spherical aberration but not chromatic aberration. Whiteside expresses bafflement at this equation; although a third-degree equation needs to be solved in order to determine the relevant curvature, it is not this one. See Newton, *Mathematical Papers,* 3:554–555.

Chapter 6

1. Key to the success of this experiment is that the second aperture be near enough to the prism so that the light reaching it will still be white (the light refracted by the prism does not produce a complete spectrum on a screen located very close to the prism, because it takes a certain distance for the different color-producing rays to be separated from one another). Unlike the direct light of the sun, in which all the different kinds of rays are understood as mixed together indiscriminately along each and every ray path in the beam, the already refracted light has undergone a "sorting" according to angle. Although the refracted light close to the prism is still basically white, it is a white produced by rays that are no longer traveling along parallel paths. Because violet rays are refracted most sharply, red the least, and so forth, the violet rays that ultimately pass through the narrow slit

at H must strike the first face of the prism at o, slightly below the spot where the blue ones that pass through the slit strike it (at n), which strike the first face below the green rays that pass through H (which fall on m), etc. By moving a wire or other long, narrow obstacle up and down at the right of the prism one could thus block out the red (at k), yellow (at l), green (at m), blue (at n), or violet (at o) component of the spectrum. (Note that in the figure the rays are pictured as undergoing an inexplicable change of direction upward at slit H.)

2. Seventeenth- and eighteenth-century modificationists did not have an explanation, but nineteenth-century versions of the wave theory did, in terms of the propagation and interference of secondary waves in the medium.

3. They were not observed until the first decade of the nineteenth century. For them to be visible it is necessary to use very narrow rectangular apertures. As we have just seen, Newton did occasionally use such apertures, but since he was not looking for this inconspicuous phenomenon, since it is not clear that his experimental setup was ideally suited to producing it, and since in general Newton tended to work with circular apertures, it is not surprising that he did not discover these dark absorption lines.

4. For additional considerations, see C. V. Raman, *The Physiology of Vision*, esp. 33–41.

5. Newton was not the first to notice the different behavior of refracted light—Johannes Marcus Marci (1595–1667) preceded him in this—but he was the first to exploit it theoretically.

6. See C. L. Hardin, *Color for Philosophers*, esp. 1–58.

7. See John Locke, *An Essay on the Principles of Human Understanding*, Book 2, Chapter 8, Section 9–10. Locke lists the primary qualities of bodies as "solidity, extension, figure, and mobility" in Section 9. In Section 23 he gives them as "bulk, figure, number, situation, and motion or rest."

8. This means that sensation is understood as the transmission of information about the thing rather than the integral transport of some of its characteristics. Put in these terms, it becomes an open question again whether the Aristotelian understanding does not have some merit after all: "information" implies that it is not the messenger or medium that counts but rather the way a message shapes or forms the recipient.

9. A different compromise, one that underlies the present tuning of notes on a piano keyboard, *equal temperament*, allows small deviations from the mathematically perfect ratios for all intervals except the octave; this results in the essential equivalence of all the major scales, regardless of which note they begin with.

10. See Newton, *Optical Papers*, 1:537–549.

11. See Penelope Gouk, "The Harmonic Roots of Newtonian Science," in Fauvel et al. (eds.), *Let Newton Be!,* 118–119. It has been argued, however, that the reason Newton added orange and indigo was that they are clearly visible at high levels of brightness. In particular, when one refracts bright sunlight into a darkened room, one is able to discriminate orange between red and yellow and a dark blue between the violet and bright blue (or cyan), colors not visible at lower brightness levels. For the case of indigo, see H. L. Armstrong, "Comment on Newton's Inclusion of Indigo in the Spectrum," *American Journal of Physics,* 40:1709, 1972.

12. These are, respectively, yellow arsenic (trisulfide of arsenic), an otherwise unidentified purple pigment, the blue pigment of smalt (a deep blue glass powder, tinted by a cobalt oxide), and verdigris (a green or grayish green carbonate of copper).

13. But this principle is not simply true. In accordance with it (and the powder experiment), one might expect that if the radiation reflected from a white and a gray object is identical they will appear the same. Yet experiments can be set up showing that if a "naturally" white object and a "naturally" gray one reflect the same kinds and quantities of light in the same field of vision, the white will still appear white and the gray gray. Even a very dark gray illuminated brightly and compared to a white illuminated dimly (so that considerably more light comes from the dark gray) will appear to be a gray while the white will appear white, unless the difference in illumination is made extremely large or dazzling as in Newton's experiment. Similar phenomena can be exhibited for chromatic hues: two differently colored objects can be made to reflect to the eye identical radiation, yet they will both appear in their "natural" colors. See Edwin H. Land, "The Retinex Theory of Color Vision," *Scientific American* Dec. 1977:108–128.

14. In all four original editions of the *Opticks* this fifth fraction in the series is given as $1/16$, but that is inconsistent with the symmetry of Newton's preferred scale of just intonation. These fractions are the factors by which a string needs to be progressively reduced in length in order to sound the notes of the rising scale, in accordance with Newton's theory of the analogy between the spectrum and the diatonic scale. If the original string is 360 units long, reducing it first by $1/9$ (i.e., 40 units) gives 320 as the length of the string that sounds the whole note above the note corresponding to the 360-unit-long string; reducing this 320-unit length now by $1/16$ of its length ($1/16 \times 320 = 20$) yields a 300-unit-long string to sound the next note, which is a half tone above the preceding one; a further $1/10$ reduction (30) gives 270 units (up a whole tone); $1/9$ reduction of that (30), 240 units (up a whole tone); another $1/10$ (24) yields 216 units (up a whole tone); $1/16$ of that (13.5) reduces it to 202.5 (up a half tone); a final $1/9$ reduction (22.5) gives a string 180 units long (up a whole tone), which is exactly half the length of the

original string and so one octave above it. If you take the string lengths in successive pairs you get $360/320 = 9/8$, $320/300 = 16/15$, $300/270 = 10/9$, $270/240 = 9/8$, $240/216 = 10/9$, $216/202.5 = 16/15$, and $202.5/180 = 9/8$, that is, the successive proportions of the lengths are identical to the proportions in Newton's preferred scale (the white notes from D to D an octave above). The length 360 corresponds to extreme violet, 180 to extreme red. For precisely these numbers see Newton, *Optical Lectures,* 545.

According to Newton's instructions, the circle should be divided not so that the successive arc-intervals are reduced in length by the proportions $1/9$, $1/16$, $1/10$, etc., but so that DE is to EF as $1/9$ is to $1/16$, EF is to FG as $1/16$ is to $1/10$, etc. Divided in this way, DE (red) should take up 60.76°, EF (orange) 34.18°, FG (yellow) 54.68°, GA (green) 60.76°, AB (blue) 54.68°, BC (indigo) 34.18°, and CD (violet) 60.76°. Fig. 26 quite closely reproduces Newton's original figure, in which DE is actually about 57°, EF 37°, FG 58°, GA 54°, AB 56°, BC 39°, CD 59°; that is, the intervals are roughly those just indicated. If the divisions were arranged according to the analogy to the musical scale, as in Fig. 25, DE would be 45°, EF 27°, FG 48°, GA 60°, AB 60°, BC 40°, CD 80° (note that in this circle the boundary between green and blue is 180° opposite the boundary between violet and red). Why Newton used the proportions differently in Figs. 25 and 26 is unclear.

15. The second, third, and fourth editions all have "CA" instead of "GA." The first edition's "GH" is also mistaken.

16. For a two-dimensional figure, the center of gravity is the point where the figure can be balanced (e.g., on the head of a pin). The center of gravity of an arc of a circle is its midpoint, of a circular disk its center.

17. The C.I.E. chromaticity diagram has a quite different use. Three monochromatic lights are specified, say a certain blue, red, and green. Each point on the diagram represents a different hue, and its position encodes the relative amounts of the three lights needed to reproduce that hue. Like Newton's circle, it is based on a closed figure with the perimeter representing monochromatic light, although about one quarter of the perimeter represents the purples, which do not appear in Newton's diagram.

18. As has become evident in the centuries since Newton, however, the physics of rays or radiation is inadequate for understanding color perception. The structure and functioning of the eye, nervous system, and brain are just as important and lead to phenomena that cannot be explained by any simple psychophysical correlation of kinds of rays to perceived color. For example, colors perceived by applying pressure to the corner of the eye reveal a fundamental organization and lawfulness of the physiology and function of the eye that has little to do with specific wavelengths.

19. In Discourse 8 of the essay "Of Meteors," which follows the *Discourse on Method.*

20. For a detailed account justifying these calculations, consult Newton, *Optical Lectures*, 1:35–56.

Chapter 7

1. Newton's claim that these phenomena are difficult but also irrelevant to establishing the properties of refraction and color raises interesting questions about whether the sequence in which a science takes up problems is arbitrary. Thin-film colors are far easier to treat according to a wave conception than according to particles.

2. Fig. 29 gives an adequate idea of the look of a conchoid, which rises in the center and trails off symmetrically on both sides. For the technically minded, the conchoid is defined in polar coordinates (r, Θ) as the graph of an equation of the type $r = a + b/\cos\Theta$.

3. The object glass, or objective, is the lens in a telescope (or microscope) that is closer to the object being viewed.

4. The general formula relating focal length f to the radii of curvature r_1 and r_2 and the index of refraction n is $1/f = (n - 1)(1/r_1 - 1/r_2)$. Since the two curves in Newton's lens are equal but opposite in direction, $r_1 = -r_2$, so this becomes $1/f = (n-1)(2/r_1)$. Since $f = 83.4$ and $n = 17/11$, r_1 is almost 91 inches, and thus the diameter is about 182 inches.

5. The secant of an angle Θ is equal to the inverse of the cosine, $1/\cos\Theta$. A number x is the *arithmetic* mean proportional between a and b when $x = (a + b)/2$, whereas it is the *geometric* mean proportional when $x^2 = ab$. Thus 5 is the arithmetic mean between 1 and 9, whereas 3 is the geometric mean between them. In calling for the first of 106 arithmetic mean proportionals, Newton is dividing the difference between the two sine values into 106 equal parts and taking as the relevant sine value the entire lesser sine (that is, the line segment that represents this sine) plus 105 of the equal 106th proportionals, that is, the total length of the larger sine segment reduced by one proportional part.

Chapter 8

1. Reflexibility was already at issue in experiments 9 and 10 of Book I, Part I, where reflection of some rays *and* transmission of others took place at the second refracting face of a prism when the light was incident at very close to the critical angle (there the phenomenon was used to support the idea of differential refrangibility). Because the rays toward the violet end of the spectrum have lower indices of refraction from glass into air, they are totally internally reflected at an angle at which rays closer to the red end of the spectrum are still transmitted.

2. In the second edition of the *Principia* (1713), Newton suggested that if all the solid matter in the earth were compressed so as to eliminate pores,

that matter would occupy no more space than a nut. What ordinarily keeps the particles apart are short-range forces. The uniformity of refractive and other optical effects at interfaces suggests to Newton that they are due not to solid matter as such but to the forces between the bits of matter. For example, if bodies consist of small, solid parts, then no matter how smoothly you polish a surface it will remain uneven at a microscopic level, and a perfect mirror will be impossible. If the optical interface is really a uniform force field between extraordinarily tiny particles, however, then the field might easily affect rays in a sufficiently regular way to produce perfect mirrors, dense but transparent solids, and the like.

3. The last two sentences are not asserting that this proposition is wrong or extremely doubtful but that (as the following sentences explain) it must be taken in a somewhat loose rather than a very strict sense.

4. Most theories had explained transparency by the existence in bodies of pores through which light could pass unobstructed.

Chapter 9

1. The first edition of the *Opticks* (1704) included Queries 1–16, which are chiefly about light, the forces that bodies exert on it, the connection between heat and light, and the effect of rays on eye and nerves as the impressions made by the rays are conveyed to the sensorium. The first Latin edition (1706) added seven more Queries, which chiefly discuss the notion of an ether; these were renumbered 25–31 in the second English edition (1717), and another eight Queries (17–24) were inserted after the original sixteen.

2. This is a case in which Newton's figure is more accurate than his description.

3. The fourth English edition introduces the misprint "flexibity."

4. This idea is not entirely original; the ancient Stoic philosophers began the long psychological and medical tradition that postulated the existence within the nerves of a very fine, highly elastic substance (called *animal spirits*) that served as a medium for effecting sensation and motion. A not dissimilar notion is found in Descartes. Yet Newton can be credited as among the first to try to unify such a rigorous theoretical and experimental investigation of light with the physiology of vision.

5. For a lucid account of the principles of the theory, see Sabra, *Theories of Light,* 198–230. The hypotheses were that the ordinary ray is produced by spherical waves transmitted solely in the ether, whereas the extraordinary ray was transmitted by spheroidal or elliptical waves transmitted by both the ether and the particles of the crystal.

6. For discussions see Manuel, *A Portrait of Isaac Newton,* and, by the same author, *The Religion of Isaac Newton.*

7. These two notions, that light was a pressure without actual motion and that it was transmitted instantaneously to infinite distances, were associated with Descartes's theory of light.

8. Huygens, of course, had shown how a light wave motion could go around corners only to be canceled by the combined effects of the secondary wavelets, a fact that Newton does not mention.

9. Newton conceives of the emission of light particles from shining bodies as a process of very rapid acceleration by a repulsive force.

10. Newton's point here is not merely that the acknowledgment of a power or force is preferable to hypotheses about some unverified mechanism of how a solid can be vaporized but also that none of the hypothetical mechanisms can possibly explain such an expansion in volume without assigning incredible properties to the particles (e.g., that they are hoop- or spiral-like entities, and that the diameter of any individual particle can vary at least a hundredfold [a hundredfold increase in diameter would result in a $100^3 = 1,000,000$-fold increase in volume]).

11. Newton is describing here a tendency for motions in the universe to slow down (unlike Descartes, for whom the total quantity of motion in the universe was constant). If there is such a tendency, there needs to be some source or sources that will produce new impulses to keep the motion going. Toward the end of the paragraph Newton presents several sources of activating force: gravitation, fermentation, the natural production of heat and light, which all counteract the "entropy" or decay of the usual motions of bodies.

12. This is directed against Descartes's theory that each star is surrounded by a vortex of circulating matter and that in general motion takes place by means of vortex flows.

13. Newton is arguing that the ultimate components of matter are not subject to deterioration or fracture, because if they were the whole universe would wear out.

14. Newton is dividing the task of science into three phases: (1) the identification of the qualities or properties of things as manifested by phenomena and experiments; (2) the discovery of certain general laws that are able to account for the regular appearance of these qualities; (3) the discovery of the causes of these laws. The *Opticks* has taken as its tasks chiefly (1) and (2), although with hypotheses about the ether and fundamental forces Newton is beginning (3) as well. Yet he is also asserting that (1) and (2) by themselves already amount to "a very great step in philosophy."

15. Again there is the shadow of Descartes, who suggested in his cosmology that God had created the world according to laws such that even if the original state of matter was chaotic there would have gradually emerged a world like the one we experience.

16. Newton's God is thus everywhere present in nature, or, more accurately, nature is totally present to God, so that he knows it through and through and can effect in it any changes he wills.

17. It is instructive to compare these methodological prescriptions to the "Rules of Reasoning in Philosophy" at the beginning of Book III of the *Principia,* 398–400.

18. Newton is referring to the end of Book I, Part II, where from Proposition VII to the concluding Proposition XI he explains how the properties of light he has discovered enable us to give accounts of all the colors produced by light, in particular prismatic spectra, the rainbow, the "natural" colors of bodies, and the recomposition of decomposed white light.

Chapter 10

1. For example, in the middle of the eighteenth century Moses Harris devised an improved color circle that more accurately reflected the complementary relationships of the hues 180° apart, and in the first decade of the nineteenth century the German painter Phillipp Otto Runge devised a color sphere that embraced saturation, lightness, and hue.

2. See Wood, "The Nature of White Light," in *Physical Optics,* 648–666, and Jenkins and White, *Fundamentals of Optics,* 249–250.

3. See Hardin, *Color for Philosophers,* Chapters 1 and 2.

4. A *telescope* (thus not a single compound lens but a lens system) with the near elimination of chromatic aberration had already been produced in 1733 by Chester More Hall. The total elimination of chromatic aberration for all wavelengths is not possible.

5. It is also known that in the period of his early optical studies Newton did fairly extensive experimental work with compound lenses consisting of two glass lenses enclosing a space filled with water or other liquids and that he had recognized that chromatic aberration could be reduced, if not eliminated.

6. It is a tribute to Newton's accuracy in measurement that in making the case for waves around 1800 Thomas Young was able to rely on the accuracy of the data in Book II of the *Opticks.* The twentieth century's reconception of light as sometimes particle-like, sometimes wave-like would seem to rehabilitate a Newton-like strategy of appealing to particles subject to periodic fits. But the twentieth century's understanding of light is quite different. Newton thought that light consisted of particles with fixed characteristics that could also set off cyclical or wave-like accessory phenomena (namely, the fits). Twentieth-century theory, to the contrary, argues that photons, as quantum entities, cannot have all characteristics predetermined and that in some situations the photons have to be understood as intrinsically wave-like (because interpreting them as particles leads to contradictions of both experiment and theory).

Select Bibliography

The books and articles listed here, most of them cited in the notes, are just a sampling of what is available on Newton and his optics. Students who are interested in knowing more about Newton's life and the full range and background of his interests might begin with the essays in Fauvel et al. (eds.), *Let Newton Be!,* and then proceed to the long but richly detailed biography by Westfall, *Never at Rest,* or his shorter *Life of Isaac Newton,* and to the selection of short pieces by Newton in *Isaac Newton's Papers and Letters on Natural Philosophy.* (The latter volume includes very good introductions to the different subject areas by leading scholars; of particular interest to readers of this guide is Thomas Kuhn's introduction to the 1670s letters on optics. The three books by Frank Manuel emphasize Newton's psychological biography and his theological and religious concerns. Newton's alchemical work is examined in Dobbs, *The Foundations of Newton's Alchemy.* The philosophical and scientific background to Newton's early work and to his philosophy of nature is nicely presented in the introductory essays of McGuire and Tamny; good histories of seventeenth-century physical science that prominently feature Newton's contributions are Westfall, *The Construction of Modern Science,* and Cohen, *The Birth of a New Physics.* For broader interpretations of the significance of the seventeenth-century scientific revolution and Newton's role in it, one can consult Cohen, *The Newtonian Revolution,* and the arguments of Bechler in *Newton's Physics.*

An indispensable reference work for Newton is Gjertsen, *The Newton Handbook.* Bibliographical information up to 1975 can be found in Wallis and Wallis, *Newton and Newtoniana;* for the period since, there are the annual bibliographies in the history of science published by the journal *Isis.*

The most comprehensive source for optics before 1600 is Lindberg, *Theories of Vision;* for the seventeenth century, it is Sabra, *Theories of Light.* Optics after Newton is the subject of Buchwald, *The Rise of the Wave Theory of Light.* Jenkins and White, *Fundamentals of Optics,* is a clear and thorough modern optics textbook; there are many others like it. On questions of color, one might begin with Hardin, *Color for Philosophers;* Wasserman, *Color Vision;* Rossotti, *Colour;* and, in a more technical vein, Nassau, *The Physics and Chemistry of Color.* Newton's optical work continues to be a subject of lively interest; one might begin with some of the works listed here by Shapiro, Westfall, Lohne, Bechler, Guerlac, Sabra, and Laymon.

For biographical information about other figures named in this book the reader might profitably turn to the *Dictionary of Scientific Biography;* although with the passage of time its bibliographical information will need updating, it is the single best source for brief lives of scientists and natural philosophers.

Primary Sources: Newton's Writings

The Correspondence of Isaac Newton. H. W. Turnbull, ed. (vols. 1–3), J. F. Scott, ed. (vol. 4), and A. R. Hall and L. Tilling, eds. (vols. 5–7). 7 vols. Cambridge: Cambridge University Press for the Royal Society, 1959–1976.

Isaac Newton's Papers and Letters on Natural Philosophy and Related Documents, 2d ed. I. B. Cohen, ed. Cambridge, Mass.: Harvard University Press, 1978.

The Mathematical Papers of Isaac Newton. D. T. Whiteside, ed. 8 vols. Cambridge: Cambridge University Press, 1967–1984.

The Optical Papers of Isaac Newton, vol. 1, *The Optical Lectures 1670–1672.* Alan E. Shapiro, ed. Cambridge: Cambridge University Press, 1984.

Opticks. New York: Dover Publications, 1952.

Opticks: or, a Treatise of the Reflexions, Refractions, Inflexions and Colours of Light. 1st ed., London, 1704; 2d ed. with additions, London, 1717; 3d ed., corrected, London, 1721; 4th ed., corrected, London, 1730.

Sir Isaac Newton's Mathematical Principles of Natural Philosophy and His System of the World. Trans. A. Motte, rev. F. Cajori. 2 vols. Berkeley and Los Angeles: University of California Press, 1934.

Secondary and Other Sources

Bacon, Francis. *The New Organon and Related Writings.* F. H. Anderson, ed. Indianapolis, Ind.: Library of Liberal Arts, Bobbs-Merrill, 1960.

Bechler, Zev. "Newton's 1672 Optical Controversies: A Study in the Grammar of Scientific Dissent." In *The Interaction between Science and Philosophy,* Y. Elkana, ed., 115–142. Atlantic Highlands, N.J.: Humanities Press, 1974.

————. *Newton's Physics and the Conceptual Structure of the Scientific Revolution.* Boston Studies in the Philosophy of Science, 127. Dordrecht, Holland: Kluwer Academic Publishers, 1991.

Boyer, Carl B. *The Rainbow: From Myth to Mathematics.* Princeton, N.J.: Princeton University Press, 1987.

Bricker, Phillip, and R. I. G. Hughes, eds. *Philosophical Perspectives on Newtonian Science.* Cambridge, Mass.: Bradford Books, MIT Press, under the auspices of The Center for the History and Philosophy of Science of The Johns Hopkins University, 1990.

Buchwald, Jed Z. *The Rise of the Wave Theory of Light: Optical Theory and Experiment in the Early Nineteenth Century.* Chicago: University of Chicago Press, 1989.

Cohen, I. B. *Franklin and Newton: An Inquiry into Speculative Newtonian Experimental Science and Franklin's Work in Electricity as an Example Thereof.* Memoirs of the American Philosophical Society, vol. 43. Philadelphia: American Philosophical Society, 1956.

————. *The Newtonian Revolution.* Cambridge: Cambridge University Press, 1980.

————. *The Birth of a New Physics,* rev. ed. New York: W. W. Norton, 1985.

Descartes, René. *Discourse on Method, Optics, Geometry, and Meteorology.* Trans. P. J. Olscamp. Indianapolis, Ind.: Bobbs-Merrill, 1965.

Dictionary of Scientific Biography. C. C. Gillispie, ed. 16 vols. New York: Charles Scribner's Sons, under the auspices of the American Council of Learned Societies, 1970–1980.

Dobbs, Betty Jo Teeter. *The Foundations of Newton's Alchemy: or, "The Hunting of the Greene Lyon."* Cambridge: Cambridge University Press, 1975.

Fauvel, John, Raymond Flood, Michael Shortland, and Robin Wilson, eds. *Let Newton Be! A New Perspective on His Life and Works.* Oxford: Oxford University Press, 1988.

Gjertsen, Derek. *The Newton Handbook.* London: Routledge & Kegan Paul, 1986.

Guerlac, Henry. "Can We Date Newton's Early Optical Experiments?" *Isis* 74:74–80, 1983.

————. "Can There Be Colors in the Dark? Physical Color Theory before Newton." *Journal of the History of Ideas* 47:3–20, 1986.

Hardin, C. L. *Color for Philosophers: Unweaving the Rainbow.* Indianapolis, Ind.: Hackett Publishing, 1988.

Hooke, Robert. *Micrographia, or Some Physiological Descriptions of Minute Bodies Made by Magnifying Glasses, with Observations and*

Inquiries Thereupon. London: Martyn & Alestry, 1665; reprint New York: Dover, 1961.

Huygens, Christiaan. *Treatise on Light*. Trans. S. P. Thompson. In *Great Books of the Western World, vol. 34. Newton. Huygens*. Chicago: Encyclopaedia Britannica, 1952.

Jenkins, Francis A., and Harvey E. White. *Fundamentals of Optics*, 4th ed. New York: McGraw-Hill, 1976.

Laymon, Ronald. "Newton's *Experimentum Crucis* and the Logic of Idealization and Theory Refutation." *Studies in History and Philosophy of Science* 9:51–77, 1978.

Lindberg, David C. *Theories of Vision from Al-Kindi to Kepler*. Chicago: University of Chicago Press, 1976.

Lohne, Johannes. "Experimentum Crucis." *Notes and Records of the Royal Society of London* 23:169–199, 1968.

Manuel, Frank E. *Isaac Newton, Historian*. Cambridge, Mass.: Harvard University Press, 1963.

————. *A Portrait of Isaac Newton*. Cambridge, Mass.: Belknap Press for Harvard University Press, 1968.

————. *The Religion of Isaac Newton*. Oxford: Oxford University Press, 1974.

McGuire, J. E., and Martin Tamny. *Certain Philosophical Questions: Newton's Trinity Notebook*. Cambridge: Cambridge University Press, 1983.

Nassau, Kurt. *The Physics and Chemistry of Color: The Fifteen Causes of Color*. New York: John Wiley & Sons, 1983.

Raman, C. V. *The Physiology of Vision*. Bangalore: Indian Academy of Sciences, 1968.

Roberts, Michael, and E. R. Thomas. *Newton and the Origin of Colours: A Study of One of the Earliest Examples of Scientific Method*. London: G. Bell & Sons, 1934.

Rossotti, Hazel. *Colour: Why the World Isn't Grey*. Princeton, N.J.: Princeton University Press, 1983.

Sabra, A. I. *Theories of Light from Descartes to Newton*. London: Oldbourne, 1967.

Scheurer, P. B., and G. Debrock, eds. *Newton's Scientific and Philosophical Legacy*. International Archives of the History of Ideas, 123. Dordrecht, Holland: Kluwer Academic Publishers, 1988.

Sepper, Dennis L. *Goethe contra Newton: Polemics and the Project for a New Science of Color*. Cambridge: Cambridge University Press, 1988.

Shapiro, Alan E. "Kinematic Optics: A Study of the Wave Theory of Light in the Seventeenth Century." *Archive for History of Exact Sciences* 11: 134–266, 1973.

———. "Newton's Definition of a Light Ray and the Diffusion Theories of Chromatic Dispersion." *Isis* 66:194–210, 1975.

———. "The Evolving Structure of Newton's Theory of White Light and Color." *Isis* 70:211–235, 1980.

Smith, A. Mark. "Getting the Big Picture in Perspectivist Optics." *Isis* 72:568–589, 1981.

Steffens, Henry John. *The Development of Newtonian Optics in England.* New York: Science History Publications, 1977.

Wallis, Peter, and Ruth Wallis. *Newton and Newtoniana, 1672–1975: A Bibliography.* Folkestone, Kent, England: Dawson, 1977.

Wasserman, Gerald S. *Color Vision: An Historical Introduction.* New York: John Wiley & Sons, 1978.

Richard S. Westfall. *The Construction of Modern Science: Mechanisms and Mechanics.* New York: John Wiley & Sons, 1971.

———. *Never at Rest: A Biography of Isaac Newton.* Cambridge: Cambridge University Press, 1980.

———. *The Life of Isaac Newton.* Cambridge: Cambridge University Press, 1993.

Wood, Robert W. "The Nature of White Light." Chapter 23 in *Physical Optics,* 2d ed. New York: Macmillan, 1921.

Index

About the Author

Dennis L. Sepper is associate professor of philosophy at the University of Dallas, in Irving, Texas. He is the author of *Goethe contra Newton* (1988) and has published numerous articles on modern philosophy and science. Currently he is working at a book on imagination in Descartes's thought and developing an Aristotelian–hermeneutic philosophy of the politics and ethics of science.